AMERICAN
DREAM

DISCIPLINE, RESILIENCE, ENDURANCE, ADAPTABILITY, AND MENTORSHIP TO SUCCEED AND WIN IN LIFE

MAGDA KHALIFA

Publishing services provided by  Archangel Ink

ISBN | paperback: 978-1-7340539-1-3

DEDICATION

To everyone murdered on September 11, 2001, and to the warriors
who have died on the battlefield since

To the first responders, warriors, and contractors who have died
from invisible wounds, toxic exposures, and polypharmacy

You are not forgotten.

ACKNOWLEDGEMENTS

I thank the Lord for giving me strength through my journey, and for guiding me when I was blind. To our 45th President, Donald J. Trump, whose example of servant leadership, and doing the right thing for love of our great country has inspired me greatly. To all the brethren I served with, too many to name; I am honored to have served with you all of you, and pray you all find your purpose and peace in life.

To Grant and Elena Cardone, for your vision, and for the impactful work you do everyday to inspire and help so many people build their empires, to find purpose and to build a better life for themselves and their families.

To Pete Vargas, for helping me realize my value to the world, during my very raw moment of vulnerability. You are gifted more than words can describe, and I am ever grateful to you.

To Silent Warrior Foundation and Task Force Dagger Foundation, for caring, and for helping me when I needed it. To JB and DB, for being family to me, and for being there.

To the legendary KH, JH, and CT, my life has changed entirely because of you. I thank you from the bottom of my heart. I would not be where I am today if it were not for your efforts. To D&T and JM, for helping me in way I needed most when I needed it most. I remain forever grateful. Love you guys so much!! To CT, a special thank you for believing in me and seeing something I couldn't see. It was a very vulnerable time; your belief in me gave me the strength I needed to grow instead of fall. You are one of the most genuine and caring people I know, and I hope to live a life as honorable as you.

To the amazing BT, JT, and KG, for your servant leadership, and everything you've done, and continue to do. Ever grateful! Love and appreciate you guys endlessly!!!! To CPT H., for your leadership and for everything you taught me to be a better Soldier. To Keith & Keline David, and to Geoff Dardia for what you have created, and the help you've provided me, and so many of us. You are really making a difference!

To JB, MS, BL, CS, JQ, JS, and SS - for everything over the years. Let's do this!!! To my awesome military brothers, who made the time to help me on this project: Jonathan W., Pat F., Doug Patteson, tom@2aholster.com, Greg Chabot, Scott, and Dave West. I greatly appreciate you all! To James, the first real friend when I returned from war, who was there for me when I needed a friend in my darkest days.

To Megan, for believing in me, and for pushing me, literally! - into growth in a way only you could do. To Boone Cutler, for giving me purpose and a mission when I really needed one. To DN, for the inspiration and heartfelt words that helped me push through! To Whitney Harstad, for loving me for who I am, even when I wasn't at my best, and for your unconditional friendship and support over the years. You are a genuine friend and I appreciate you. To Les Page and Sonya "Red Sonya" Leibowitz, for pouring into me to become a better motorcycle rider, and all the memories on two wheels!

To Matt Feliciano and Ian Sattler, for your genuine support and friendship through this journey. To the great bosses and colleagues I had over the years, who I learned so much from and enjoyed working with immensely - Tom Veneroso, Ben Barnhill, JG, and AG. To Leona B. Hunt, BA, copy editor, leonahuntproofreading.com, for going above and beyond. You are an amazing soul! To the fabulous Kristie Lynn, for helping me greatly with production! xo

CONTENTS

INTRODUCTION

To whom much is given, much will be required.
–Luke 12:48

America's caught a cold, as evident in the pockets of radicalism that have erupted throughout the nation. The good news is the root cause of this malaise is curable. We have become a nation of zombies with so many Americans on psychotropic medications that are controlling their lives. I've figured out some things on my journey and as a result, have been successful in avoiding prescriptions altogether. So it's on me to help others do the same. If fewer people were under the influence of psychotropic drugs or addicted to opioids, the clearer the collective thinking, meaning less than optimal conditions for extreme idealism to spread.

Like many of you, I have not mastered anything nor accomplished anything extraordinary. I am not a world champion athlete, a world leader or a billionaire - yet. But I have done something less than 1% of the US population has done - which has intrinsic value and perspective worth sharing.

I'm a protector and a fighter who cares deeply about people. I have been very fortunate in my life and have a desire to pay it forward in a big way. I can't sit on the sidelines and **do nothing** while anti-American voices **within** America spew unthinkable things, incite racism, and disrespect our flag and those who have died for it. They disrespect our police, our rule of law, and push socialist agendas to young, impressionable, confused minds, making our Founding Fathers roll over in their graves, not 20 years since Americans stood united against evil on one of our country's darkest days. Nor can I sit on the sidelines and enjoy the life I designed and conveniently ignore the problems – that's not in my DNA.

The worst feeling in the world is watching something terrible happen and not being able to do anything. I'm grateful I can do something - in fact, I can do a lot. This book is a starting point to help those who need and seek help. It's much more than a turn-of-the-century story from the battlefield to business.

What I can do is lead by example, sharing my story and providing value, solutions, and a way ahead for those that seek guidance in a noisy, confusing, conflicted world - **especially** at a time when our economy is so robust, abundance is everywhere for the taking, and being in business has never been easier. It is my duty to share. Half a century is right around the corner for me, and I realize that it is high time to give back to the world in a way that I uniquely can, lest I take what life has taught me to the grave.

If you take nothing else away from this book, know that you have the power to heal yourself. You can absolutely have access to the recesses of your brain to uncover underlying issues and unprocessed thoughts that hold you back from becoming the best version of you and from reaching your potential and beyond.

Why "DREAM"?

DREAM is an acronym to remember. It stands for the dream elements: **D**iscipline, **R**esilience, **E**ndurance, **A**daptability, and **M**entorship. The D.R.E.A. is vastly missing from America today. Not enough people have developed the Discipline, Resilience, Endurance and Adaptability in life to face the challenges of the future - or of today!

With the average American living into their late 70s, it is important to learn the characteristics necessary to succeed and win in this thing called *life*. You can have an average life and live it peacefully, and that is fine for many people. Congratulations, you've succeeded.

For those who crave more, and want to win so they can leave a lasting legacy - for their families, future generations, communities, country and society overall - it is important to develop Discipline, Resilience, Endurance, and Adaptability. It is also essential to recognize the value of Mentorship - to help guide you to where you want to go. You develop the former and *choose* Mentorship. If you already have the characteristics but are missing Mentorship, consider adding it to your life and see how much quicker you get to where you want to go. We should all be striving to win in life by being better than the person we were yesterday. Mentorship will help you connect the dots from being capable to taking it to the next level and winning.

In this book, I share my stories and the lessons I learned, and how I developed these traits, so you understand it's not an overnight development. Shout out to Endurance! I explain how mentors took me to the next level - so you can get there as well, in less time than it took for me.

I wrote this with the intent of providing answers, hope or a nugget that can help change or transform your life. People are confused today and need strong, grounded people of integrity to provide life guidance. History has been rewritten, and it is hard to think straight because of technology and information overload, or because prescribed substances affecting thinking. Critical thinking is waning in the over-stimulating sound bite world we live in.

I wrote this as we approach 2020 as a Gen Xer, caught between the baby boomer and millennial generations. I am old enough to remember the significance of old-school values, which are slowly disappearing - and young and adaptive enough to recognize important trends and the importance of change. I serve as a "generation-bridge" because I care deeply about people, society, and humanity overall.

The reality of the collective American dream may be evolving, but what truly makes America great is timeless, despite the conditions of any given era. You **can** navigate life's setbacks with intestinal fortitude. This is a book for champions in the making. For winners who love winning, and will not settle for less than their calling.

One thing is for certain: you need a bulletproof brain, and a bulletproof mindset, to move from darkness or pain, toward success and significance. This takes drive, commitment, and dedication to grow - and the elements of DREAM to achieve this.

Bringing old-school values back

This extreme nuttiness is a phase. American values always reign supreme; that's in our country's DNA. However, it's incumbent upon many of us who remember life before the information age took over to rise up and lead - like we are capable of doing, lest we let anti-American influence fester. You know who you are.

The world has changed greatly with the rise of the information age and 4th- and 5th-generation warfare. The traditional post-World War II baby boomer generation defined their collective American dream with a home with a white picket fence, and a career or job with a retirement pension - "40 hours for 40 years, for 40% of the paycheck". The millennial generation is taking over the declining baby boomer generation. Millennials are not buying homes or getting married as their baby boomer counterparts of the same age did. They seek mobility and experiences instead of "getting settled".

The advent of technology, global inter-connectedness, and social media has greatly influenced generational change. A young person today does not have to follow the traditional college-to-work path that worked so well for previous generations. There are so many non-traditional business opportunities and new pathways toward wealth, as we approach 2020.

This book is bringing old-school back: self-reliance, courage, service, tenacity, duty, hard work, hustle and a seemingly lost value: integrity. I'm not perfect, far from it, as you will soon read. But I've become better. And you can too.

I welcome those who want to be part of my tribe - those who have integrity, courage, and genuinely want to be part of a movement to enrich and better their lives, and the lives of those around them and those who need them. I open my tribe to those who are lost, who have no direction or guidance, but have the burning desire to grow and be part of something bigger than just themselves.

The standard in America has fallen, and "kinder, gentler" has morphed into weaker, confused, incapable, codependent, and inoperable. The thing is, it is still very cool and very much a blessing and a gift to live the American dream, be self-reliant, work hard, thrive in a capitalistic environment, desire great success and significance, and greatly serve others because you can – regardless of race, color, religion, gender, sexual preference, or origin. Patriotism has always dominated, even when tested by radicals. It's the American way and like titanium, it's only getting stronger. My life reflects those titanium properties, and yours can too. We're in this together. Let's raise the bar!

F is for fixed. F is for fail.

First thing is first: Success in anything in life comes down to your outlook. Are you growth minded or fixed minded? Being growth minded is essential to achieving time freedom, health freedom, and mind freedom. Growth minded people are always learning and open to change. Fixed minded people have set beliefs and are resistant to change… or growth.

If you are a taker who doesn't want to change because you are comfortable, and personal change and growth is uncomfortable to you, I can't help you. No one can.

Everyone in America has a chance to do anything. I empathize if you live in an inner city where conditions are not optimal, but that's an excuse. That's when you have to look deep inside and ask yourself if you are going to be part of the system and the environment you live in, or are you going to move on? You are probably wondering, How do I move on? Here's the thing: God helps those who help themselves by opening doors for those who look for them.

If you are an adult and believe your life sucks, it's because of the decisions you've made or your perspective that it sucks. There is nothing stopping you from changing that. And it's never too late to change.

Perhaps you don't have a safe route to get to a clean, decent and quiet library to collect your thoughts in silence and develop your game plan. If you are resourceful and open-minded, you can always find a way. When you dwell and focus on something, it becomes attainable, but you have to be clear on that one thing you're trying to get to. Trust me, if you obsess over it, it will happen. The universe will allow it to happen. Truly, you really do have that power within you. But you have to want to use it. We have all heard "If you want things to change, you have to change." How badly do you want it? Maybe you are comfortable, and that's okay; however, be real with yourself first. If you've already adopted a fixed mindset, you've already failed.

Throughout my life I was presented with information that was very much the opposite of what I knew and believed for many years. And because of the new knowledge I was presented with, I then changed my behavior (which I discuss in this book). That is indicative of a growth mindset. People don't like to change at a certain point in their lives. It's much easier for them to stick with the status quo. They have a fixed mindset. Now they don't need to embrace or change their behavior overnight; I'm not suggesting that. However, running away from knowledge or the truth simply shows a fixed mindset.

A person's mindset is often reflected in their overall attitude. Fixed mindsets lean towards negative. I've learned to accept people for who they are. I'm not going to stop caring about people with a fixed mindset. But I limit my association with them. I choose to spend the majority of my time with people who are growth minded.

> Write down the first ten people in your life that come to your mind. Based upon what you have read in this paragraph, determine whether each person has a growth mindset or a fixed mindset, and circle them - *G* for growth or *F* for fixed.
>
> G / F _____   G / F _____
>
> G / F _____   G / F _____
>
> G / F _____   G / F _____
>
> G / F _____   G / F _____
>
> G / F _____   G / F _____

Do you have more growth minded or fixed minded people influencing you? Remember the adage, "If you are the strongest/smartest/most successful/fastest/richest (choose one) person in the room, then you are in the wrong room." If you are the most growth minded person in the room, it may be time to find a new room.

Everybody's propensity to achieve, succeed and win sustainably in life, versus episodically, is incumbent upon the mindset they are made of - a fixed mindset or a growth mindset.

The power and difference found in this book

A lot of people speak about mindset, and habits and such, and have great messages, but sometimes they are merely reiterating lessons they learned, not lived or experienced firsthand. It's not a terrible thing; we need more positive narratives and guidance in the world, regardless. However, it is a bit hard for me to connect on anything more than a superficial level with someone who can talk the talk - but hasn't walked the walk. Again, there is value in their speaking about great things, for sure! But are they the storyteller, or the story?

This work is written as a memoir to give context to the lessons in this book. Ideally these lessons are understood and absorbed better because they are coming from real experiences over a lifetime rather than simply coming from great words.

My book's mission is to help you grow and take action in your life, from having learned something new or through seeing things through a different perspective. This book is not for everybody. You may not be ready to receive the message. Do not be discouraged; it may not be the right time for you. Timing is important. Come back to it later, and it may make more sense when the time is right.

I've taken these steps and they have not failed me throughout life. I will show you my journey of becoming "superhuman" - which enabled me to transcend traditional mental limits through nutrition - the natural way. With the addition of mentors, I have learned to refine the process to achieve success quicker. I share with those who truly wish to make a difference in this world, starting with their own lives.

A true mentor will always tell you what you need to hear, not what you want to hear. I intend to pay it forwards and do the same for you in these pages, like others have done for me.

In a land of abundance, there is no excuse not to grow like bamboo, get out of your comfort zone and advance society. Be a lion, not a sheeple! Lions have the means to fight the evils in civilization. I challenge you to rise as a leader, and I will help you learn how to **lead yourself,** at your pace, to become your best version and make a real difference in your own way for your loved ones, for your community and for the world.

We will start off with a simple question: What were you put here on Earth to do?

_____

_____

_____

THE GOLDEN HOUR

Show me how it ends
It's alright...
–Breaking Benjamin, "So Cold"[1]

It was a long Fourth of July weekend in 2007, all alone in the new apartment, and it felt weird. I felt like I could sleep forever. I was lethargic and had headaches that would come and go, but I didn't know which hurt worse - my head or my heart. I thought of the guys I was deployed with who were still overseas in the fray. How were they holding up? The news trickled in only so much. I had some emails from my friends I had left behind when our unit rotated back, but understandably they could not say much. I felt vulnerable thinking, If something happened to me right then and there, how many days would it be before anyone knew? I was not in communication with my family. No surprise, they sided with the Marine in the divorce. I had no support anywhere. I was separated from my tribe and did not know how to function.

With my mind in such a volatile state, my thoughts continued. No one really knew where I was living. And no one really cared. I had just started the job, and if I didn't show for a few days, they probably would have assumed I had quit. I concluded that eventually a rotting body would produce a stench in the hallway, and after a few days,

---

1   Breaking Benjamin, "So Cold," track #1 on *So Cold*, Hollywood Records, 2004, LP.

maintenance would come in. Death was very much on my mind after seeing so many variations of it the past year - firsthand.

Thoughts like these ran through my mind as I lay there on the carpeted floor watching the assault and tracer rounds light up the night sky - a.k.a. fireworks in the distance in Norfolk. My mind wandered to a few weeks ago, climbing the sandbags on the side of my trailer in the middle of the night to get to the rooftop and watch the opening salvo of 5-73 CAVs (5th Squadron, 73rd Cavalry Regiment) assault on As Sadah, a few kilometers away, across the flat desert. While I waited for H-hour with my night vision goggles, I noted the quiet on the forward operating base (FOB) and looked over the rows of trailers where most of the soldiers were sleeping. Someone tossed a ChemLight on the way back from the portajohn. I couldn't sleep that night; I was so excited to witness history before our team joined them in the field for the next phase of the operation.

I sat up and looked toward the kitchen counter and saw my purchase from earlier in the day. I bought a bottle of Jack Daniels, but I'm not sure why, as I didn't drink whiskey. I looked at that bottle of alcohol on the marble surface and was compelled to drink it all, knowing while it would not necessarily solve anything, it would temporarily reduce the pain I was in - the pain in my heart and in my head. That moment was a golden hour for me. As I looked at it, I also knew that the minute I took that first drink from it, given the state I was in, I would be pressing quit on life. I would be turning toward something unnatural to escape my pain.

Three decades of something - what, I don't know - forced me to push through the night and not quit. I anguished all night - tossed, turned, sobbed and had vivid flashbacks that made no sense - until eventually, I passed out. I remember waking up feeling like shit. I went to the kitchen in this tiny apartment and saw the bottle on the counter. My head was throbbing, but seeing it sitting there untouched was a small win for me. I had the mental toughness to make it through one of the roughest nights since I was back, all alone with my thoughts and no crutch. I knew right then I could make it through any night and just keep going.

GROWING UP JERSEY

Life started rather ordinarily for me. I was born in the '70s and raised in north New Jersey by my parents who were legal immigrants. Mom left her native Colombia in her 20s for a better life in the United States. She was one of six children who moved from city to city because her father was a Captain in the Colombian Navy.

She arrived in New York City in 1968 and started working clerical jobs while attending classes to learn English. She went to the Armed Forces Recruiting Station in Times Square - at the height of the Vietnam War - but was told her English was not good enough to join the Army.

In 1972, Mom, a blue-eyed beauty with long hair down to her waist, went dancing at the YMCA on Lexington Avenue. She met a quiet, well-dressed man who asked her to dance.

"Where are you from?" she asked.

"Egypt," said the man.

"Oh, wow! I never met a mummy before!" my mom replied. It was love at first sight. They married in a small ceremony on Park Avenue, and I was born a year and a day later.

Dad hailed from a family of doctors and engineers. His father, Abdel-Mones Khalifa, had studied mechanical engineering in England, graduating from Sheffield University in 1939. He then worked at the air force academy in Egypt, writing the book *Aeronautical Engineering* - the first book on the topic written in Arabic.

Dad was the eldest of six and had an adventurous and entrepreneurial spirit as a kid, building things and identifying problems in his community to solve to make money. He studied abroad in Germany as part of the student exchange program and graduated from the University of Cairo with a degree in engineering. He had a good life in Egypt but wanted more; he saw the United States as the land of opportunity.

I was a shy, quiet kid. I loved music and reading books - especially biographies of historical figures. My brother and I participated in a lot of sports and activities outside of school. While I was enrolled in dancing, karate, softball, and ice skating, I was hardly a natural athlete and had to work hard at anything physical I engaged in. Dance was the only one I could do decently and easily. As a kid I did not enjoy failing miserably at all these activities. It probably would have helped if I wore my glasses for my 20/70 vision, but I didn't, so I kept failing.

Mom was a great nurturer and would drive us around to all of our practices and games, as Dad commuted to the Bronx where he worked as a director for an engineering firm. The company he worked for did industrial work for the US Air Force and was involved in repairing damages in one of the Twin Towers after its first terrorist attack in 1993. I have early memories of him coming home from work, and we would sit and eat dinner, that Mom prepared, as a family, and then he would leave to New Jersey Institute of Technology to study in a master's program. He eventually attained two master's degrees in engineering and computer science.

He was super smart, and when I needed help with math or calculus homework, I would marvel at how he would merely look up, think for a moment and blurt out the answer. Dad was an innovator as well and held patents for a self-locking device for outdoor enclosures and for a ceiling panel assembly. I didn't understand what they were but knew enough to recognize and respect that he was smart and accomplished.

Mom, Dad, and I. 1974.

We lived in a modest home and grew up middle-class. I have great memories of Dad taking me to see the Yankees play at the iconic Yankee Stadium! I collected baseball cards and idolized these great athletes. At the baseball games, I enjoyed my first taste of beer and delicious Yankee franks. My favorite part of the game was watching players stealing bases. Though it involved skill to successfully complete, it was a quick, decisive action that involved risk but could help the team, and advance the objective of the game - winning. Not surprisingly, one of my favorite players was Rickey Henderson, a base-stealing king.

STEALING BASES

Life is a game, with an unknown and finite number of innings. Stealing bases in life will help you get more wins. More wins keep you going in life. Every time you take action and grow as a person, you are stealing a base and increasing your chance of getting to home plate for a win. Once you stop growing, life ends. You become the spectator in the seats, and are no longer the all-star in your game of life.

Time is a valuable commodity you never get back. Once it's gone, it's gone. There is no short cut to hard work but you can start stealing bases in life by saving time. For example, outsourcing tasks will buy you more time. You may love to cook and love to drive, but when you are busy you can open an app and outsource these tasks, gaining valuable minutes to focus on goals and dreams. It may not be as glorious as stealing home plate, but 45 minutes of time in a set 24-hour day is invaluable.

Family memories

My childhood was happy; we traveled all over the country and overseas for vacations.

Road trips were great! We would all sing along to an eight-track cassette of Johnny Cash, in Dad's blue '72 Skylark.

I have fond memories of fishing with my dad in the Finger Lakes region of upstate New York, going to Disney World, and driving along the beautiful Pacific Coast Highway in California. We visited warships, battle sites and military museums around the US, which I always found interesting, as I loved history.

At home we loved watching black-and-white documentaries about World War II, the *Wings of War* series on Discovery Channel, westerns, and any movie with Clint Eastwood or Marilyn Monroe. I loved the sound of the WWII era's warplanes and machine gunfire.

Holidays were always fun and festive. We would barbeque and watch fireworks at the local park for the Fourth of July; and Mom would cook a feast for us for Thanksgiving. She made many of our costumes for Halloween, and Christmas was always merry, with music playing, laughter, everyone dancing, and a beautiful and festive tree with the voice of Santa indicating a trove of presents had arrived. When I hear Bing Crosby or *Feliz Navidad*, I remember those memories. One day I found a cassette tape with my dad's voice saying,

"Ho! Ho! Ho!... Ho! Ho! Ho!" My parents made sure our experience was magical.

My dad didn't read to us, but he would tell us stories. My favorite were the adventures of siblings Jean and Michael and their motorcycle. They would ride together all over the country on adventures. Always a good time!

In front of the iconic steel beams inside the World Trade Center. 1978.

Seventh grade

It's funny how you forget about things from so long ago because they have little or no bearing on your life decades later. However, when you consciously think back to periods of time in your life, details emerge from your memories.

This is one such story. I was one of the youngest in my class year and also the runt. I had metal braces for three years and my clothes were not designer fashion. Like in most school systems, there was a group of popular kids that hung out. Renee was the tough girl in the group. For some reason, she started picking on me in the hallways between classes. I don't remember what she said; others said it as well, but she was the most persistent. I guess they got their kicks from taunting their targets. This went on for a while. Then one day while we were going up the stairs to the next class, Renee was right behind me saying her usual taunting and how she was going to kick my ass.

Enough was enough. My inner honey badger awoke. I turned around to her and said, "Do you want to fight?" I gave her no time to reconsider, as I even threw out a suggestion for the location - the school parking lot. Though my heart was pounding, I enjoyed a sense of empowerment for taking ownership of this ongoing situation. I did not give a care what anyone thought or would think; I was tired of the taunting and believed this would end it - even if to my detriment. So I showed up. She showed up. And because word spread faster than a viral tweet travels today, the entire seventh grade class showed up as well.

It was a sunny day, and somehow we ended up on the street adjacent to the school's playground. Renee was one of the bigger, more athletic kids in the class. This was going to suck, but I was committed to defending my honor. So there we were on the street facing each other. I am not sure how many kids showed up to see me get my ass kicked versus how many showed up to see what Renee would deliver. I remember looking her in the eyes. Thinking back now, she was probably more talk than walk and was not sure how to go about this beatdown.

She grabbed me with both hands, and then I did the same to her. We were in a deadlock, and I remember shaking from the effort. I don't quite remember what happened next, except all the noise coming from the crowd of school kids around us. I distinctly remember Johnny Mack stepping forth and putting his ice cream cone in my hair. The crowd was noisy, some laughing, some yelling. Were punches thrown? I literally cannot remember. I believe there were a few shallow punches but not more than that. A bloody brawl, this was not.

My English teacher arrived at the scene and tore us apart. Then the principal's office contacted both of our parents. I wasn't in trouble with my parents, but there certainly was no coddling involved. Thank goodness. Soon both of our parents had a face-to-face meeting with us kids in the room. Some sort of peace was established on behalf of the parents' intervention. After that incident, no one bullied me anymore. I was still a misfit but I had stood my ground and faced the problem. Owned it!

This battle was nothing compared to the much tougher battles yet to come. It was just one survival incident that didn't really define my life. However, thinking about it now for the first time in many, many years, I am glad that twelve-year-old Magda was principled enough at a young age to stand up for herself in the face of danger. Looking back, I suppose this gave me a sense of conviction in self, even if the odds were against me or if nobody understood where I was coming from and I stood alone. Noted. Self-reliance for the win.

BE SELF-RELIANT

Damaged people are dangerous. They know they can survive.
–Josephine Hart

Deep thresholds are forged over time through experiences, increasing personal capacity, and capability. A person who is coachable can learn skills. At some point, despite mastery of skills, internal fortitude will be tested. Are you fixed minded, waiting for someone else to solve your problems? Or, are you growth minded and intent on taking ownership and action?

You can't rely on anyone else to build you. They can support you, but where do you stand when they are not around? When you are all alone? You have to look inward for your strength and build that solid foundation first. In doing so, you will learn who you are at your core and how much you want to survive. For those who have existed in the alone space for some time, it's candy when any kind of genuine support comes your way.

People have asked me: Aren't you afraid it's not going to work out? How do you know doing that was the right thing to do? It's this simple. First and foremost, I trust in myself. I **know** what I have done; I **know** what I can do because I have been tested repeatedly and persevered. I know what my threshold is. So failing does not scare me, as I've failed many times before. But I've always had resilience to keep going. Become the person you trust the most. When you do, you exude a quiet confidence because your internal compass is calibrated and locked on.

If you want to increase your threshold and self-reliance, do this: when everything is going right, plan for the next hurdle by giving yourself a new challenge. Rise to a state of being unstoppable. Not in a hashtag kind of way, but in reality. Live a life where you have enough conviction in yourself that any underestimation of you is at the detriment of anyone who does so.

The right people are attracted to that inner strength. Others may not see it. Whatever. It does not matter. You and you alone are responsible for the decisions you make and where you go and what you do in life. No one gets it right all the time. What matters is that you keep going. If you have a strong sense of self, strong sense of self-worth, and a strong history of overcoming and adapting, you become a force to be reckoned with. You're not going to build yourself up that way by playing it safe and staying inside a comfort bubble.

Work = money = freedom

Home was an environment of learning, growth and responsibility. And as such, I helped my brother with his newspaper delivery route. We were up very early before school delivering in the rain/snow/sleet/hail. Come to think of it, Mom would drive us around sometimes in the ole' station wagon. I didn't think about it until now, but I can appreciate how she was leading from the front, teaching us the value of hard work and team work at a young age.

Dad used to sit and "lecture" us. Of course a kid doesn't want to be lectured, but I remember vividly many of the takeaways of what he said. He said, "You kids are so lucky to be born in this country. I didn't come here until I was in my thirties. If I was born here, I would be a millionaire, maybe a billionaire by now. You were born here. You can be anything you want to be. Anything at all." And this is why I have always felt empowered.

I was as curious about everything as Dad was, and we bonded over learning together. He subscribed to numerous publications like *Smithsonian, National Geographic, Forbes* and the *Wall Street Journal*. I learned fundamentals of financial literacy at home and developed habits of fiduciary responsibility at an early age. Mom co-sponsored a credit card for me at the local clothing store, so I could start establishing credit when I was 16. She said when you make a dollar, automatically put half of it in the bank and live off of the rest so you don't live above your means.

When Dad was home in the evenings, the TV was turned to the news or financial channels. Then for entertainment we enjoyed MTV, westerns, James Bond movies, and the World Wrestling Federation's "Saturday Night Main Event." Dad was always sharing articles he read with us kids. I can't say I was particularly interested in everything he was sharing at the time, but it did impress me that there was a lot of things in the world to learn about. I remember one article that had a graph showing what would happen if you started investing the max allowed in an IRA each year, starting at age 18. It showed that through compounding interest, you would have a million dollars by the time you retire. That sounded like a lot to me as a kid. I couldn't wait to be 18 and start contributing to my future.

In the meantime, I started working whatever I could whenever I could. The Jersey hustle was real, and work meant money. Money meant freedom. In those years, it was freedom to buy cassette tapes of my favorite rock 'n' roll bands, like Guns N' Roses and Ozzy Osbourne. It was also freedom to be out in the world because as I was getting older, I noticed my parents were becoming strict with me - protecting me, of course - but to a pre-pubescent teen who wanted to experience more of life, I saw it as if they were holding me back. But if my parents weren't strict with me, I probably would not have been as self-driven to find solutions to my problems because I would not have had problems to solve! I enjoyed the rewards of working: new grown-up experiences - working with adults, new friendships with my coworkers, and money! Glorious green!!

The summer before high school, I got a job making pizzas at the Wallington roller rink for $3.35/hr. while the DJ blasted Metallica and hip hop. Limited hours, of course, since I was 13. Work was fun! I liked work! I worked anything I could to earn money, even babysitting. I was thrilled to have earned $500 that summer. That bought several cassette tapes, and I put some money in the bank too.

Growing up Jersey, I learned early on how to build relationships with everyone. It's what we did. There were no barriers; you could dislike someone and they could dislike you, but you both knew where each other stood. And you could still do business together! That kind of rapport building goes a lot further than pretentious banter, which is limited. I was fortunate to learn early in life that everything falls on relationships! People prefer to do business with people they like, know and trust. So even if they don't like you, but they know and trust you, you can still make a deal together. You can be the most affable, likeable person, but if they don't know you - if you have no relationship - collaboration of any kind will be hard-pressed.

High school

With high school came new friends and experiences. I joined cross country, winter track and spring track. It felt good to have found a sport I could participate in and not suck entirely, even if I was one of the slower runners. I enjoyed going to practice and spending time with my fellow team mates. I loved the runner's high I would get running the three-mile route in Garret Mountain in Paterson, listening to the long songs on Metallica's *And Justice for All* album on my Sony Walkman cassette player. I used to draw on my notebooks at school and loved drawing the Metallica logo with the trademark lightning bolts on both the first and last letter; I would draw my name in the same style, since it began with *M* and ended in *A*.

For training, we ran longer distances to build up our endurance. I didn't realize it at the time, but the physical endurance built up my mental endurance and stamina that would pay off later in life in many ways. One of my best friends Jenn Heslin was into metal music too. After track practice, we would blast Iron Maiden in her room across the town park and watch the lightning strike the trees in the distance. We read metal magazines like *Kerrang!* and played heavy metal records backwards on the turn table to hear the hidden messages, ha! That was our social media equivalent way of connecting with people in 1987. No drugs, just happy, healthy metalheads running track, throwing shot put and riding our bicycles.

Heavy metal music became my "drug". On July 20, 1989, I attended my first metal concert, Metallica, with The Cult opening up. My friends and I sat in the nose-bleed section, but it didn't matter. Metallica was loud, played hard and rocked! On the stage was a giant

Statue of Liberty, blindfolded. She fell dramatically in the encore, breaking into pieces.

In November 1990, we did a family trip out West. I remember sitting in the back seat as we drove from Arizona to Las Vegas. I was excited that we were going to visit the Hard Rock Café to celebrate my birthday! As we drove through the dark desert, the news on the radio was talking about the situation in Middle East and how the US may be going after Saddam Hussein in Iraq. In January, I remember watching the war unfold on CNN. It was exciting and victorious and had really piqued my interest. Footage of bombs dropping and stuff blowing up fascinated me.

I went to my first Slayer show in 1991. The original lineup played with Megadeth, Anthrax, Public Enemy and Alice in Chains in the Clash of the Titans tour. My friend and I had floor seats in the iconic Madison Square Garden. It was a wild show, with a crazy mosh pit and M-80 explosions going off from the balconies and the crowd ripping seats up. It was so intense that lead singer, Tom Araya, had to tell the crowd they wouldn't be able to come back and play in New York City again if it kept going like that. Little did I know at 18 years old that Slayer's music would be **the** constant in my life over the next three decades, the music's intensity being the source of motivation and strength to get through my toughest moments - be it grueling workouts or mental pain during my darkest days.

High school went by quickly. I got an electric guitar and started taking lessons. I even joined a glam rock band but was fired rather quickly because I sucked and only knew a few chords. We played in my parents' garage but since we needed money to play in a real studio, we got jobs at a local fair and were part of the magic show at the freak show tent. We were sawed in thirds, played the role of exorcist girl with the spinning head, and breathed live fire - just like Gene Simmons of KISS.

I remember the carnies getting excited when two men came to

check out our show. I had no idea who they were but heard their names were "Penn and Teller", and they were legendary in the magic world. The channel USA filmed my exorcist girl act for a break-bumper Penn was hosting on the network's show "Up All Night." Between acts I was on a stage in front of the tent with an eight-foot Burmese python wrapped around my shoulders, enticing people to enter the exhibit. I couldn't believe they were paying me to do this. I racked up 14 hours a day and it paid cash. We probably were working more hours than child labor laws would allow – who knows?

Mom and Dad came by one of the last days of the fair. I don't think Dad was pleased to see how I was dressed – nothing too scandalous – I was in character in heels and miniskirt. But I was a good kid, working hard and making money, and I know he valued that hustle – money was survival.

Hustle and grind

I had a lot of summer jobs as a kid – telemarketing was by far the worst! Who wants to be that guy bugging people in their homes and getting hung up on? At 15, I worked as a cashier at Roy Rogers but was not keen on coming home smelling like a greasy biscuit. And I used to love their biscuits, too. My hair absorbed the smells of the fast food joint - gross. Working there was invaluable because it made me realize I didn't want to work fast food, so I needed to **hustle harder**. I eventually found a job at the local deli. Per law, you couldn't use the slicing machine until you were 16, so I could only work the cashier. My boss was way cool and let me play the epic hard rock station WSOU 89.5 FM on the radio quite loudly. I guess he figured people were only going to be in there for a New York minute.

Dad always said, "Become invaluable at work and you will always have a job." I showed up early, did not take days off and enjoyed whatever job I was doing. In my senior year of high school, I started working in the mailroom for a company that did printing and micro-film services. I could see the people working the printers down the hall. Many of them commuted in from New York and caught the bus back together.

Joe, the supervisor, was a Vietnam Veteran. Because he stood out and there was something different about him, the way he moved and how he looked at things, I paid attention to him. He didn't say much, but I got the sense that this man was in control of everything that went on in the entire department, even as he stayed in place and manned the large printer. He was very... aware. Very stoic. Others went out of their way to avoid him. It was interesting to watch inter-actions. I befriended the adults I worked with and couldn't wait to be done with high school so I could get out in the real world. There seemed like there was a lot to learn and experience, and school felt limited. Adults were so much more interesting.

I moved upward from envelope stuffing in the basement to administrative assistant in the office for the tech team of program-mers. My boss, Tom, was great to work for. He included me in their daily morning meetings where they discussed technical stuff. It was a very diverse group, with someone from Thailand, a Sikh, a Russian, an Indian, a Venezuelan and several others. Every month we would go to dinner at a different person's favorite ethnic restaurant, which enriched my life. Most of them were at my wedding years later, and I've kept in touch with a few almost 30 years later.

Ever the hustler, I worked as an extra for various commercials and films, which was a great way to make $100 for a day's work. I also worked part-time security at the Meadowlands Sports complex. It was great to be part of a team, and though standing long hours in inclement weather at Gate D during the football games kind of sucked – as did dodging spitting Jets fans, - it was offset by some of the cool experiences like being right next to the stage while the Rolling Stones played, hanging out with the team, and drinking beer that we confiscated at the gate after our shifts. The team felt like a family and they looked out for one another. I remember Michelle, the sweetest soul who stripped at the local club, who always bought me a hot chocolate on the coldest winter days. We were the only females on the security team so were looked out for by the guys, most of who were in their 40s and 50s. We all showed up and worked our part, but chivalry was not dead in tribal New Jersey.

In the summer, we would run through the tall weeded fields during concerts, chasing the illegal shirt vendors. Once I got separated from my teammate but felt victorious when I caught one shirt vendor by myself and he handed over his stash to me. Running track had paid off. It was a strange cycle. He was hustling too. Money moved and favored those who chased it hardest: entrepreneurs. Growing up Jersey, you see everything. You do everything. I saw a lot there.

I got posted to the wives' lounge for the Devils' home games. That was a very easy, chill gig, and since I loved hockey - and the Devils - I got to watch the games on the TV in the hall. Very little work involved. And I was being paid for this! I took pride in my post and took my job seriously, feeling the need to protect the families inside the room from strays and creepers. They were kind and gave me a beautiful gift for Christmas in appreciation for my work. That was unexpected and so thoughtful. The Devils were world champs and these families were on fire, and they still took time to remember the little people like myself. Class acts.

Time management and priorities

Since I just wanted to work and earn money, I did not want to go to college. However, Dad convinced me to go to state college and major in business administration so I could apply it broadly to a career. He used to say "Communications is the future, and with business, you can do anything."

I started commuting to college at 17 while working several part-time jobs. Since I needed wheels and had responsibility and savings from work, my dad agreed to subsidize 33% of my first vehicle in order for me to get something reliable. He helped me buy a dependable used car to keep me safe on the roads while I drove through the streets of Newark and Jersey City at odd hours going to school and work. I took pride in ownership since it was my first major purchase. I also paid my share of the insurance, which was on their policy, and learned the importance of regular maintenance.

Trust had been extended to me, and I was very willing to meet my end of the deal. It was the continuation of a steady pattern of personal fiscal responsibility. I started maxing out an IRA annually as soon as I turned 18. I've always been grateful to my parents for instilling that sense of responsibility early on. Later on I worked for companies that offered employee match on 401k contributions, so it was a no-brainer to max my contributions. The nest egg had been planted and it kept growing. I didn't know it at the time, but the discipline I had to build wealth the old way set me up for big gains when combined with the newer ways of creating wealth I learned about later on.

I had no choice than to become highly disciplined and efficient at time management so I could get studies in and make time to continue working part-time jobs so I had money. There were no excuses. I knew what I had to do to advance my life and knew no one was going to do it for me. If it was meant to be, it was up to me and no one and nothing was going to get in my way. Little did I know at the time that I was already executing the "whatever it takes" mindset I learned from one of my mentors years later.

I adapted to the circumstances, sacrificed social time and made it happen. Some classmates at school didn't understand why I chose to work hard instead of joining a sorority or attending campus parties. I never felt the need to explain and conversely wondered why they felt the overwhelming need to be liked and to fit in through groupthink and predefined social acceptance. Didn't we all leave that behind in high school?

There is nothing wrong with social groups, fraternities and whatnot. We are humans, and it is natural to want to connect with people. However, I suggest you place more emphasis on the activities that will help you achieve your goals and desired end state first. If your goal is simply to fit in and feel accepted because your scope is narrow, I believe you will get great value out of the coming chapters.

DO "YOU"

We start out in life caring too much about other people's opinions. Sure, if they house and feed you or employ you, they have a say on some things. To that I suggest, why not start making your own money earlier on so you have freedom?!

Stop worrying about what other people think about you. They are going to judge you regardless, so why waste energy on that? Are you setting your goals for yourself, or are you seeking recognition from others? We all have the freedom to define our individuality, a privilege denied to many other people worldwide who are forced to conform. Value that freedom. You will never become unleashed and unstoppable in life if you are constantly stopped short of pursuing what you want to do because of how others may judge you.

Army appetizer

After my sophomore year, I transferred to Seton Hall University because I wanted to be part of their hard rock FM radio station, 89.5 WSOU. I had DJ'd at Jersey City State College's WGKR station and loved being a part of the music scene. I picked up a concentration on management information systems, so I combined business *and* communications, as Dad advised.

I received a flyer in the mail about Seton Hall's ROTC program and went in to the campus to meet the leadership. Observing some of the tall, squared-away cadets who had been to Ranger school, I was totally intrigued by how they stood differently and carried themselves, even among other cadets in the program. I was not a tall, studly athletic type, but I wanted to grow as a person and wanted more in life, and this all this intrigued me greatly. So I said, "Hell, yes."

That summer of 1993, I shipped out to Cadet Challenge in Ft. Knox, Kentucky for six weeks. It was a blast! I loved everything we did, like ruck marching, rappeling from a tower, land navigation, and learning troop leading procedures.

Our First Sergeant (1SG) was a silver-haired Vietnam vet who nicknamed me "Qaddafi" because he couldn't say Khalifa. I remember opening up an MRE (Meal, Ready to Eat) when we were in the field and the wrapper on the M&Ms had the Olympics logo from 1984 in Los Angeles. It seemed peculiar that my food had been preserved for almost ten years.

I loved the nightly live fire course where you had to low crawl under concertina wire while tracers shot right above you. The best was when we did night patrols, moving around with MILES (multiple integrated laser engagement system) gear on our weapons. As we approached the OPFOR's (opposition forces) camp, the sound of Metallica could be heard coming from one of the tents – which got me even more pumped up to capture them! I won't say the experience was easy, because keeping up physically on the runs and rucks with so many tall cadets was challenging, but it was fun, and hustlin' at everything in life was my jam.

I grew as a young person in many ways, built confidence in myself in all kinds of areas that were not part of my Jersey upbringing, and learned leadership from the best leadership organization on the planet – priceless. Overcoming each struggle became a win. Seeing others struggle with whatever helped me realize I could do this and even brought out a little competitiveness in me, as leadership wasn't just sheer physical effort. I could control professionalism, duty, motivation, and merely showing up at the right time in the right place in the right uniform. Everything else I could learn or improve on. It felt good to win.

A lot of the cadets were stressed out about whether they would be awarded the college scholarship upon completing training because they couldn't afford college. I had been diligently working for a few years and was able to pay my way (albeit with ten years of student loans to pay off), so I didn't have the pressure they did and I enjoyed the experience. In the fall I was sent a letter saying that I was awarded the scholarship contingent upon continuation of military science courses in college and acceptance of a commission as an officer upon graduation from college. Now I was down for it and eager to go to Airborne and Air Assault schools the following summer as part of the continuation of cadet training, but Dad did not believe serving in the military would be ideal for me. He talked me out of commissioning, telling me it wouldn't be a lucrative career for me. Very reluctantly, I trusted his guidance. I got a taste of the military and loved it, but for now I was working on getting that piece of paper so I could get out in the real world full-time and have real freedom. I continued to read many books about Vietnam, now having a little context of what the Army was about. War fascinated me.

Continuing the grind

My father encouraged entrepreneurship and shared information with me about phone cards that were in high demand. I started a business distributing these Megacards to retail businesses. I would drop off these cards on consignment at gas stations, convenience stores, and strip clubs. I didn't realize it at the time, but I was probably enabling the drug dealers throughout Newark and Passaic counties. Dad was right, communications was a hot commodity.

The store managers would call me up for more, and I would stop by to pick up the money for them and resupply. I planned my route so that it was aligned with my commute to school. The biz was **hot** for a few months, until large global corporations started producing their own phone cards and undercut my small business with price and broader distribution.

I learned about a program at the University of Hawaii where you could take college courses over the summer and have room and food included for six weeks for a thousand dollars. Aloha! I presented the idea to my dad with the proposition that knocking out some credits would free up more hours for me to work during the regular semester. He was always open-minded and supportive about travel opportunities, and this combined education as well.

So instead of jumping out of planes like I originally wanted to, I went to Hawaii. The campus was beautiful, and I met many new friends from all over. I tried sushi for the first time and partied at Moose McGillicuddy's. As luck would have it, Anthrax was playing on the island. I went alone, put on my best Jersey tough girl look, and made my way to the front of the stage amidst the sea of very large locals and Marines. And right there in the middle of the mosh pit, I locked eyes with a Marine. It was love at first slam!

He was stationed on Kaneohe Bay. We dated the entire time I was on the magical island, hiking Sacred Falls before it was shut down, jumping off the Pali cliffs, watching the fireworks at the Aloha tower, and visiting the Punchbowl cemetery for Memorial Day services. I attended the Marine's ceremony when he was promoted to Corporal. He taught me how to drive stick shift in his little Mazda on an old runway on Bellows Air Force Station. That was fun! I returned home to my last year in college and eventually received letters in the mail from the Marine from Japan.

Graduation was fast approaching, and I could not wait to be done already. I lived for my various jobs but still managed to graduate in four years at age 21. No more school for me. That's what I told myself then. Things were becoming more complex at home. Between Mom,

Dad, and me, you had a mix of three very strong personalities and a mix of three distinct cultures, so things could be explosive. Of course I was often vetoed by the governing body of parents whose house I lived in - so I hustled hard and was driven to create my own space in life and be free of control.

DINK life

I decided to move out to Hawaii to find a job there so I could also be with the Marine who had returned to the island. I booked my flight for the day after graduation and packed a couple suitcases. Dad did not approve of my plan. Not because I was looking for work in Hawaii but because he did not approve of my living with the Marine. Technically I would be living at his platoon Sergeant and his wife's house on base. On the morning of the flight when Mom was about to take me to the airport, Dad said to me, "If you go, don't come back. You are disowned." He wouldn't even look me in the eye. It was a sucky, stressful situation all around, but hell, I was 21, had a degree and money in the bank. Time to go out and build my life.

I got to Hawaii and enjoyed reuniting with the Marine and living with his platoon Sergeant and his cool family. He gave me a ring to claim me. Smart move. But something was missing: I could not find work. This frustrated me to no end. Work was much more abundant back in the New York City area. With all due respect to the Marine who would have been content supporting us on his E-4 pay - that wasn't going to cut it for me. I was a confident fireball with dreams and goals. I couldn't even get a job as a bank teller, and though I was living off my savings and the kindness of his platoon Sergeant who housed me, this was not good.

I remained in communications with my mom, who essentially cut a deal with my dad. If the Marine and I married in New Jersey, he would accept me back in the family. And he would pay for the wedding! I knew a great deal! The Marine was getting out of Marine

Corps anyway, so the timing worked out. So basically he would just have to move back to New Jersey with me, as I wasn't keen on staying on an island I couldn't find work on, paradise as it was. It all worked out too well.

I came back to New Jersey, landed my first full-time job in New York City while the Marine transitioned out of the Corps. A wedding or getting married wasn't even anywhere on my radar as an ambitious, healthy, happy 21-year-old, but having a man in my life was. Mom and I planned a wedding in six weeks; he arrived in New Jersey, and we married, even though I forgot my bouquet as I walked down the aisle! My parents practically adopted him as a son of their own - which I knew they would. The Marine was a great guy and could easily make Mom laugh and loved working on house projects with Dad. I was glad to see them all get along so well while I focused on my career. I had started working as a recruiter for computer professionals on Wall Street, placing highly skilled technicians at financial institutions and banks like Goldman Sachs and Chase.

This was 1995 and information technology was booming. The Marine and I paid a small rent to live in my parents' attic while we started our married life together and saved for our future. He got a job working at UPS at night while going to school. I took the train from New Jersey to Hoboken, then the PATH (underwater subway) to the World Trade Center. Then I passed the giant Bugs Bunny at the Warner Brothers store, took the long escalators up to the ground level and walked a block to the office on the corner of Broadway and Liberty. On nice days, I would pay the extra 50 cents to take the ferry across the Hudson River instead of the PATH train.

Life was good and seemed very promising, but I wanted more. I saw how much the people I was placing were making - $2000 a day for SAP developers - and I wanted my piece of the pie. I started looking for entry-level computer jobs to get in on the game. One day one of the successful sales ladies in the office came over to my desk and asked if I had lunch plans. I didn't, so she invited me to lunch. I was pleasantly surprised. We walked over to a popular eatery in the area and ordered.

I was not sure why she wanted to do lunch but valued the time with her, observing everything - her fashionable hairstyle, her sharp power outfit. After some small talk, she started talking to me about how I was dressing. My skirt to the knee and basic blouse was not Wall Street enough. I was hearing her loud and clear, but felt hurt and confused because even Mom thought my outfit was nice and professional. And then, the ankle bracelet: she shook her head and said, "No ankle bracelet." She stopped just short of critiquing my hairstyle, which looking back probably should have been the first item on the list.

I thanked her for the lunch and her time and feedback, but felt like a loser the rest of the day. I wished she had been more forth-coming about it in the office, instead of inviting me to lunch and making me feel good about myself first, then embarrassed. My ego was bruised, but I knew she was doing me a huge service by having a real conversation about how life worked on Wall Street. How would I ever aspire to be the success she was if I didn't dress the part? It made sense, but I didn't know the first thing about looking all put-together - and I certainly didn't have the cash to afford several outfits that cost hundreds of dollars. Obviously, those were all excuses I told myself and believed. The truth is - though I did not know it at the time because I did not know myself enough - I didn't want to pay the price to look the part to have the job badly enough. I dropped the ankle bracelet, did the best I could with my wardrobe and breathed a sigh of relief when I got an offer for a C+ programmer position back in New Jersey.

CHECK THE EGO

Someone who is genuinely looking out for you will tell you what you need to hear, not what you want to hear. If you find someone that is being real with you, value that relationship because that kind of feedback is critical to your growth as a human. Pay it forward by doing the same for others. If you are principled, it will make a lot of sense to do so. Remember, we're all trying to survive and win in this thing called life. You are more valuable to someone by being honest and pointing out what they may not see or understand. All these years later, I am still grateful that the sales lady made the time to speak to me and tell me what I needed to hear, though I didn't appreciate it at the time. Drop the ego and grow!

Life worked out pretty well, as the Marine entered the computer field as well and landed a job near mine. I moved on from programming - as coding was called back then - to tech support, taking PCs apart and putting them together. The industry was evolving, and as such, people in the computer field needed to evolve as well. The Marine followed the Novell network administrator track, and I followed the Microsoft network administrator track. We both obtained respective certifications and worked positions often very similar for years. Salaries were great, and we enjoyed being DINKs, dual income, no kids.

We moved out of my parents' home quickly and lived in a few different apartments before buying our first home. Life was good; we worked hard and enjoyed weekends and holidays, traveling a lot and spending time with the other fun, young couples in our new neighborhood. It was nice to finally have friends who were married to hang out with, as we were babies when we married - 21 and 22. We were very active, ran marathons and went to awesome metal concert tours all over the country.

The 1998 New Year's Eve Black Sabbath reunion in Bank One Ballpark in Arizona was epic! - along with Megadeth, Pantera and

Slayerrrr! I thought New York City mosh pits were intense, but here people on the floor were lighting fires on the baseball field and jumping over it as the metal gods raged on. The retractable roof revealed an epic fireworks display – ringing in a prosperous 1999.

The late '90s were very abundant and bountiful. I drove a sweet new fire red five-speed 3000GT, and we both enjoyed riding our motorcycles in the fall to Michie Stadium in West Point, parking right next to the stadium to watch Army football. The ride back down the Palisades Interstate Parkway was beautiful. We loved to eat out at various amazing restaurants throughout Jersey and go clubbing at all the hot spots of the '90s, like Joey's in Clifton and Jenkinson's in Point Pleasant. The Marine was well ingrained in my family; Dad really, really loved him like a son. We loved to cruise and invited my parents to come along on a cruise with us.

Big hair, fast car and full of life. 1999.

I was very much a social butterfly, organizing fun get-togethers for our friends and ourselves. We both enjoyed entertaining in our home and looked forward to having everyone over for "The Sopranos" Sundays, to enjoy a nice spread of the best Italian local foods and a new episode of the show we all loved, which was filmed all over where we lived. When friends from out of town visited, I would give them a Sopranos tour, driving by the famous fake pork store in Kearny to the Bada-Bing strip club, a.k.a. Satin Dolls.

While things were going well, life would always throw challenges. When one of the companies I worked for merged with another company, employees received vested stock options per our employment contracts. Major cha-ching! Except for about a dozen of us in the tech positions who were mistakenly not included in the forced vesting. I wasted no time in hiring one of the top employment attorneys, spending tens of thousands to fight for what was contractually due to me. I declined their paltry settlement offer. Despite hard evidence supporting the case, as well as testimony on their side that indicated a cover-up, we lost in arbitration. Stunning. Soon afterward we learned that the acquiring company did not want my case to be an example for the others to come forth and sue. They ensured it did not happen.

While it was a long and costly battle, I'm glad I did it, as I learned a lot from the process, including the power of large corporations. I also learned a lot about myself and the power of being principled, of once again standing up for yourself, your dignity, your property and what you believe in. I also learned the precise sign my attorney observed in the arbitrator that told him we lost before it was official. Knowledge of this little tip served me well in later years. I started to understand the value of body language. Even then as others viewed a large battle lost, I found a small nugget. A growth minded person always finds the gold in a landfill.

I was otherwise very comfortable in life and felt stronger to face life's challenges with a partner by my side. We had our share of issues as all couples do but worked through most things. For better or worse, having money usually helped a situation because we would get a new toy or go on a trip. I loved planning amazing experiences for us. For one of his birthdays, I bought floor seats for Aerosmith - his favorite band. On another birthday, I surprised him with a visit to the Scores in New York City for a night of lap dances and champagne with three of his friends. One year we decided stay in on his birthday and order takeout, keep it low-key - as it was a Monday night, a work night, and we were getting ready to leave the following week on our dream cruise to Tahiti.

I sat in the cozy white leather chair in our fancy living room looking over activity choices for the different islands. Each island looked so beautiful! Between the flight and the length of the cruise, it was going to be a ten-day vacation, epic. I kissed the Marine, "Happy Birthday, Baby," as we went to sleep with excitement about the new travel adventure that awaited us.

THE UNTHINKABLE

The next morning was a beautiful blue-sky perfect September day - the kind that just makes you feel good! The Marine and I went about our normal routine with the TV on in the bedroom turned to the news. I glanced at it and saw the unthinkable - a plane hitting the World Trade Center. I yelled at the Marine to come watch. As I flipped through news channels with an unsettling feeling in my stomach, the Marine called his office. He worked in computer support for an insurance company, AON, and rotated between two offices - one of them located on the 101st floor of the North Tower of the World Trade Center. He knew many of the employees because he worked on their computers. Shortly after the first plane was hit, the South Tower was hit. Like millions of Americans who were watching this unfold on the TV, we knew right then this was not an accident.

Seeing was believing, and I was almost in a state of disbelief. I needed to do *something*. I could not just sit in front of the TV set or watch the skyline from my balcony about eight miles away. The Marine went to his New Jersey office, and I contemplated grabbing my Canon camera with telephoto lens and jumping on the NJ Transit train straight to Hoboken, literally right across the river from lower Manhattan for a closer look. I could be there in 15 minutes max, as the train station was nearby. Capturing some photos would certainly be useful for someone - the police, the FBI, or something. It made sense as I had a great camera and was in close proximity. Don't sit still. Do *something*.

Then that instinctual feeling set in. I felt something in my gut that sent a message through my body not to go. I struggled with this feeling and could not explain it, but then again, we don't always have an explanation for instinct. There was nothing holding me back from going but this gut feeling. I continued to watch the television as reports of the Pentagon attack came in and President Bush spoke and identified this as a terrorist attack. Then shortly after the unimaginable happened.

The South Tower collapsed.

There are no words to describe that moment. Indescribable. Time literally stopped. Like millions of people, we watched the unthinkable happen. Soon afterward, the North Tower collapsed. No words can capture that moment. Cell phones were down and for hours. Mom and I could not get through to my dad and brother who both worked in New York City. The silence of the airplane grounding in one of the busiest metro areas in the country was deafening.

I went outside to the highest vantage points I frequented, having lived in the area all my life within eyeball-line sight of the New York City skyline and those gigantic towers. As I listened to the go-to AM news radio, 1010 WINS, I could see the traffic backed up for miles on the main arteries in and out of the city. The cloud of smoke from the dual collapse dominated the skyline. I snapped some photos from a distance and reflected on how lucky I was that instinct kept me from getting closer to what would come to be known as Ground Zero. I probably would have let my unsatiated curiosity lead me onto the cross-river ferry or PATH (underwater subway) – whichever was still running in that direction - to get closer and be right there at the World Trade Center in a few minutes. To be right there as the buildings collapsed. Or maybe I would not have even made it out of the underground. This was unreal.

War in my backyard. September 11, 2001.

All day long, the news indicated bomb threats in the area. No one knew what was going to be hit next, even if planes were grounded. Mom was able to eventually communicate with my brother who lived and worked by the Empire State Building. He was going to egress out of New York City, and eventually did hours later. But we didn't yet know about my dad. The radio reported ongoing suspicious activity under the George Washington Bridge. Dad worked in the Bronx and commuted over that bridge. I prayed, as we hadn't heard from him in hours and it was anyone's guess what else the terrorists had in store. He made it home near midnight, after being diverted north to the Tappan Zee Bridge.

Our neighbors gathered in the evening in the cul-de-sac we lived in, and shared horrific stories. One of my friends came out of the PATH train into chaos and large "thud" sounds. He was disoriented and did not understand as he described a man falling and landing in front of him on the sidewalk. Oh my God! We would come to learn with the rest of the world that these were the people who were faced with the choice to jump or burn from their offices on the top floors that were hit by the jets. Every single person was grieving in their own way, and everyone was in shock.

As president of the condo board, I suggested we knock on the doors of our neighbors who we had not yet laid eyes on to do accountability and support any of their family members. Thankfully our community did not lose anyone, but communities all around us were not as lucky.

The Marine and I stayed up until 4:00 a.m. watching the coverage on the TV until sleep took over. The next few days and weeks, of course, were crazy for everyone. We tried to donate blood, but the banks were full, as there were few survivors. We both could barely eat and felt numb and in shock. We functioned, but it was through a foggy feeling of pain for what happened to the country we loved, right in our own backyard and in our Capital and in Pennsylvania next door. Countless friends and acquaintances were directly affected by the loss of life - many of their loved ones killed when the towers fell. Dad's coworker lost his son, a fireman who was saving lives in the towers.

A slight deviation from the funk I was feeling was when President Bush stood on the pile of rubble, still burning below and announced on the megaphone, "I can hear you. The rest of the world hears you. And the people - and the people who knocked these buildings down will hear all of us soon!" There was hell to pay and I could not wait to see us unleash the force of our military.

For almost a week, a cloud hovered over New York City, slowly making its way north toward midtown. The smell of rotten egg, sulfur, was pungent - even across the Hudson River, as Black Hawks (helicopters) flew overhead. From across the river, you could see clearly the shattered glass of the Winter Garden arch - blown out from the collapse of the buildings. One of the hardest things for me to see were the missing people notifications - have you seen so and so? These signs were posted all over the Hoboken terminal where so many of those killed commuted to New York. Seeing their faces gave me chills, as I realized they had walked where I was walking just a week ago. Thinking of the anguish the families were going through - brutal.

I thought how lucky I was that the Marine was not in the World Trade Center office that day, or I may very well have been posting a similar sign in hopes that someone had information on his where-abouts. Reports of hope, sounds of survivors from beneath the rubble, and stories of pets left behind in the residences around the World Trade Center come flooding in. The Marine lost 175 coworkers. He was deeply affected but pressed on working in his vital role as part of the IT department for a large corporation. The one silver lining in this American tragedy was the united display of patriotism with beautiful American flags flying everywhere - on homes and on overpasses in the highly congested metro area.

Being helpless sucked. I had no medical or construction skills to offer the efforts at Ground Zero. I called the New Jersey FBI office to see how I could help. Given my last name, they asked if I spoke Arabic. I replied, "No, but I have all of these IT skills." No luck. I called all over New York City to see if anyone could use my IT skills - not as a job but to donate my time to help in any way. No luck. I really was useless, and it was eye awakening.

One day we learned the story of Rick Rescorla, an Army veteran who fought in the infamous Ia Drang battle in Vietnam. He worked for Morgan Stanley and was killed when the towers fell as he duti-fully ushered employees out of the burning building and didn't get out in time. I turned to the Marine and said, "I know if you were there, you would not have left before everyone else in your office did." He and another Marine he worked with in the IT department exemplified the Duty of the Marine Corps long after they were out of the service - to a *T*. No doubt in my mind, they would have stayed as long as needed.

I drove south on Broadway late at night when I could finally get through the barriers and stared in disbelief as I approached the Ground Zero area and the office building where I had my first job after college. The building stood covered in white dust, as well as the Century 21 building across the street, ...then I saw the massive pile of rubble ...with workers moving all over it like ants. Good God, my city. My country. I was in shock - like so many others. Shock turned into anger.

I visited the site several times over the next few weeks and months, sometimes alone, sometimes with the Marine or with friends. We all noted how odd it was from a navigational standpoint. You used to be able to be anywhere in the city and know your bearing because of the towers which stood so tall and iconic. Now, I couldn't help but be symbolically and literally lost without them there. The fence around Trinity Church was covered with faces of the missing and memorials. Walking down Liberty Street and seeing the orange spray-painted search and rescue markings and the steel cross erected were sobering reminders that many were still missing. Somewhere in storage, I have many more prints of photos I took in the months afterward, as well as mementos -newspapers sold around the area at that time.

SEND ME

The fire in my belly was lit, and shock turned into anger. I visited the Army recruiting office in Montclair, NJ in January 2002. I wanted to deploy overseas and go "outside the wire" and not be stuck on base, so I was thinking military police. I did not meet the height requirement for military police. So the recruiter recommended civil affairs, which I hadn't heard of. This was not public affairs. He explained that civil affairs was the only reserve unit in "Special Operations Command." I was intrigued. He further explained that the job deploys soldiers during peacetime and wartime and had special requirements to get in. You would work with the best of the best and be a critical part of everything wherever you go. Boom, I was sold. Just get me there and I'll figure the rest out.

I enlisted at the Brooklyn MEPS station in March. And in May, the Marine, some friends and I rode our motorcycles to the Rolling Thunder rally in Washington, DC, supporting our POWs/MIAs. I had the smallest bike, a little 500cc Buell Blast, and the least riding experience, so they put me in the front. I taped directions to my tank and off we went. The wind atop of the Delaware Memorial Bridge was insane; it's a small wonder I didn't get blown off the bridge! I hugged the bike and throttled it to get off that bridge as fast as I could. Traffic extended the trip to six hours, but I felt like a rock star leading the pack and not getting us lost, despite the confusing beltway signs. We pulled in to a gas station to fuel up and I looked up and was taken aback by what I saw. Straight ahead was the side of the Pentagon that was hit; visible damage still showed and repair cranes still surrounded it. Looking up and seeing that was like a punch to the gut. Memorable.

The event had a record turnout - over 300,000 bikes. Patriotism was palpable among this crowd, especially months after the biggest attack on America since Pearl Harbor. As we sat in the hot parking lot for hours on our bikes awaiting our egress on the ride route, I stared at the Pentagon and thought about the attack just months prior. It fueled my desire to deploy and kick some ass, and I couldn't wait to do so. As we followed the bike convoy, helmetless, throughout Washington, DC, I enjoyed the sights, and sounds of freedom and the breeze - and thanked God I lived in such an amazing country.

Surrounded by patriots and chrome.
Rolling Thunder, Washington, D.C., 2002.

Right before leaving for basic training, I died my bleached-blond hair to brown to not stand out, as I was now serving something greater than myself. Sometimes you see the strangest signs in life. One night as I was about to go to bed, I heard a loud collision and saw my 3000 GT hit by a drunk neighbor. It was a total loss. Well, not that I was going to need a vehicle for the near future.

What makes the grass grow?

As I lay in bed in the temporary barracks, I reflected on leaving the life I had behind. I knew I was 100% committed to my next journey, and it would start the next day. It was a semi–cold-turkey departure from a rather abundant life. Life was not perfect, but I had a marriage, a community, friends, influence, a beautiful home, money in the bank, and a vibrant life full of travel, music and adventure. I couldn't fully explain to myself why I was *so* compelled to do this. But I knew I had to since 9/11 had kicked me in the gut. And the timing was right, as my country needed many of us and I was able to jump in and to do my part. Little did I know how much everything would change for me.

Transition in life is usually stressful to begin with. Transition, when you know you are leaving a good life is another thing. But alas, this was my decision and I was all in. Let's do this, I thought as I drifted off to sleep.

In basic combat training (boot camp) it was just like the books I read, just like the movies - minus owing Gunny Ermey one jelly doughnut. I found myself being a live witness to history. Not an out-of-body experience, but while I was executing the suck, my mind was thinking about how freaking cool it was to be part of US military legacy - a long line of warriors. I don't know what they put in the chow hall food, but I remember one night in the barracks at Ft. Jackson dreaming the most vivid dream of Revolutionary War Soldiers: I heard the fifes playing and saw them marching right toward me in the same barracks I was in. I could really hear the sounds. I awoke and it gave me the chills. But in a good way.

This evolution was a lot dirtier and had a lot more screaming than the Cadet Challenge I attended eight years prior. At night we would sit shining our boots, but we weren't allowed to talk during the meals. I met another Soldier who was older like me, versus the majority 17- to 20-year-old Soldiers. He had a remarkable life and career but, like so many others, was compelled to serve after 9/11. Since we had no news, radios or communications with the outside world other than snail mail (letters), I wrote the Marine and asked him to print and mail intelligence reports from a service I subscribed to. After I read them, I would pass these envelopes to the other older Soldier because he shared the same hunger for knowing what was going on in the world.

After a short Christmas break, I went to Fort Bragg, NC, home of Army special operations for advanced individual training (AIT). We trained alongside our psychological operations (PSYOP) counterparts; collectively we were the only AIT class on the post. We were trained by high-speed cadre from the US Army John F. Kennedy Special Warfare Center and School (SWCS). My inner history geek loved the significance of training there and the lineage we drew from the Army Special Forces A-teams in the Vietnam War.

Though I didn't know it at the time, it was in AIT - as we trained in the small teams, like we would operate - that I started really understanding and respecting the real reason we would fight. Sure, fighting for our country and defending our Constitution were honorable and noble causes that everyone serving believed in, but you were also doing it for the man on your left and right. I loved the sense of loyalty and family in small teams, and I remember the excitement when we were in the field and the Drill Sergeant announced that the US had started dropping bombs in Iraq! Game on!!!

Once I returned to my home unit in Staten Island, I requested to be put on the first available rosters for units deploying to the fight. As luck would have it, a unit was mobilizing at the end of the year to head to Iraq. In the meantime, I got my affairs in order and I prepared for war in a way that made most sense to me.

I've always been an early adopter when things made sense. I had been wearing contact lenses since high school but knew I couldn't do that in the desert because of the dusty sand. Not knowing about prescription goggles at the time, I thought, What would I do if I wore glasses and they fell off in the desert? I'd be useless if I couldn't fire my weapon.

So I looked into LASIK, and over the summer learned about a new technology called "Custom LASIK" from Canada that had been in the US for three weeks. I jumped in and got it done for $5,500, with both eyes being corrected to 20/15. I came home and could read the news ticker across the screen on the TV in the other room. Super powers!!! That new capability gave me even more confidence, and I was even more excited about deploying!

TAKE RISKS

Knowing your threshold allows you to take risks and decisive action. Risk leads to reward; at the minimum, it will force you to grow and will reveal your threshold. It makes your decision-making process easier when you know how much you can tolerate and how much can you stand to lose.

A simple risk formula you can apply is, "Does the value of growth outweigh the value of stagnancy?" It most often does. If things haven't changed in a long time, they probably won't change without taking risk. Don't be afraid to take a risk in life. You don't grow by playing it safe. If you fail, you learn from it. Don't delay due to fear either. Timing in risk-taking is important. As I wrote previously, "Decisive action at the 85% solution may yield steel on target versus the 95% solution that simmers for weeks and misses the golden hour. Experienced leaders know when to pull the trigger."[2]

---

2   Khalifa, Magda. "Trump Doctrine: Adaptation and Advancement of the United States in the Information Age". April 10, 2017. https://www.linkedin.com/pulse/trump-doctrine-adaptation-advancement-united-states-age-r-khalifa/.

Management of expectations

If you are looking for an impressive war story of individuals, read citations of Medal of Honor recipients and their mind-blowing acts of valor under fire and read accounts from Bataan Death March survivors. What I share with you in the next few chapters are slivers of my Iraq war *experience*, as a portion of my life experience as a whole. Think of them more as examples of leadership - or lack of leadership - in a combat setting versus a "war story." I had the privilege of serving with amazing leaders - US personnel and troops - and the unfortunate situation of serving with some shitty leaders and troops. I experienced examples of both bravery and of cowardice and experienced the brotherhood that unites against a common enemy. Like many other troops, I experienced the downside of personality conflicts that are heightened in a war setting. The war is the context, but the meat of the lessons learned are applicable in a variety of environments. I use the examples of the good, the bad, and the ugly of my experiences to drive home points that can help you navigate decisions in life - and see things through a different lens and perspective to make the right decisions for yourself, your life and your business.

When I started writing the Iraq portion, I decided to include a couple of stories that exemplified lessons learned to provide value to you, the reader. While researching an old hard drive, I found sporadic journal notes and emails I wrote that captured life at the time. It was a bit of a trip reading them 13 or 14 years later; some of these things I had forgotten about or shelved. I include some of them, edited for content and clarity - with names and units changed as necessary - because they help authentically convey the environment, the times and the lessons in this section.

Nikki

I joined the unit I was deploying with in December 2003. During the roll call I was happy to see a familiar face – another Soldier, named Nikki. We had gone through Basic and AIT training together and shared some pain and memories. We trained up with this new unit for a couple months at various locations in the US. I remember sitting in our barracks at Camp Shelby when the Sergeant Major came in and announced that the US military had captured Sadaam Hussein. Morale was very high!

We continued training and getting ready. We learned how to skip periods downrange by taking the pill back-to-back, an off-label tactic. Some of us females stocked up at the military pharmacy for a year plus supply to assure we had enough to last. More capability.

Finally we were set to head overseas in February 2004 with a brief stop in Kuwait to acclimate to the desert climate. Nikki and I had been paired up as battle buddies since we knew each other and were both from units other than the one we were deploying with. We were assigned to a CAT-A team (civil affairs team-alpha) in Charlie Company, but somehow in the confusion of the first few days of arrival to Iraq, we were split and I was channeled to headquarters in Tikrit. I was told that since I went to college and "knew computers," they wanted to keep me as a staff resource at headquarters (HQ). Oh hell, no! I did not leave my awesome life to sit at a computer in Iraq!

Thankfully they quickly figured out that Nikki did not have a female battle buddy with her, so a convoy was organized to transport me there, where I should have been from the beginning. The convoy was set to leave in a couple days. In the meantime, several of us explored the Tikrit palace. I was struck by how lush and green everything was next to the Tigris River. The battle damage from the bombing the year prior was visible with large parts of the massive solid buildings destroyed. BAM! Right before the convoy south, we went on our first mission "outside the wire." I was 100% "green": scared, nervous and excited all at the same time. Game on!

The mission was uneventful. Shortly after returning to base, we received the news: one of our teams near Baghdad was hit during a convoy. Early reports were one KIA (killed in action) and three WIA (wounded in action). My stomach sank upon hearing the news, then fury set in. I ran downstairs to the basement of the building we were in to the gym and picked up the heaviest dumbbell I could find and started pumping iron to counter the rage inside of me.

Soldiers were confused and rumors were everywhere as to whom we had lost. Then we found out. My battle buddy, Nikki, was killed by a 155 mm roadside bomb while driving the team on a mission in a soft-skinned HMMWV. My heart sank. She was 19 years old.

Nikki was the fourteenth female Soldier killed during combat operations in Afghanistan and Iraq, and the first killed in combat who served under the United States Special Operations Command (USSOCOM), formed in 1987.

Things moved quickly because the war raged on. After I was partnered with Joyce, and we were put on the manifest to immediately fly down in General Odierno's Blackhawk convoy to FOB Warhorse. He was the Commanding General of the 4th Infantry Division (4ID), to which our unit was attached.

I was asked if I would speak at Nikki's memorial service. As we flew across the desert, I scribbled everything I knew about her on my small notebook on my knee. We landed at FOB Warhorse, and Joyce and I got situated in our trailer. Since we didn't know what would happen next in our first crazy week in this country, we were all on edge. The door was locked from the inside, so Joyce crawled through the window to get us in. She was funny, and I appreciated how she lightened the mood.

Nikki's ceremony was held in the morning of February 21 on the tarmac by the hangar. I kept looking at the notes I scribbled about her.

It had to be done well for her and I didn't want to break down during the delivery. One of Charlie Company's captains came over to me. He asked if I was ready. I nodded. He must have seen the uncertainty in my eyes and pointed to the formation of Charlie Company Soldiers lined up on the tarmac. You could see the grief on their faces. Just a week ago we were all Stateside and our normal selves. The reality of war set in quickly.

"Do it for them. They need you to be strong," he said.

Noted! That's what I needed to hear. I realized quickly I needed to get over myself now and deliver a proper tribute to my fallen friend. That was the standard to achieve, that's what my battle buddies in Charlie Company needed to hear, and that is what we owed Nikki.

When it was time, I remembered the captain's words and delivered the short tribute speech, then fell back into the receiving line next to my friend from New York City, Sgt. Kevin Whelan. The blood on his desert boot was dried. Whelan received shrapnel wounds and MSG Mike Bevins lost some of his eyesight in the explosion that killed Nikki. It was a solemn ceremony, yet beauty is in everything, and I remember the beautiful voice of the public affairs Soldier who sang our National Anthem.

When General Ordierno came through the line, I had to lean my head back, as he was a very tall man. After I rendered a salute, he took my hands in his, looked me in the eye and expressed the most heartfelt condolences for the loss of my battle buddy. Though 4ID was getting ready to rotate back to the States, he said he considered Nikki part of the 4ID family and would not forget her. It was a sincere and comforting moment from one of the toughest war generals. I could see why the men he led respected him.

I carried the guilt of "Why Nikki, not me?" for a long time. All because I was deemed more valuable, fate intervened and temporarily held me back from Baqubah. Ugh.

The compound

Our team was stationed on a small, 200-meter perimeter hardened compound in the middle of the city of Baqubah in the Diyala Province of Iraq, across the street from an Iraqi police station. It was called the CMOC - Civil Military Operations Center. Baqubah was quickly becoming a volatile and congested city marred by complex and coordinated enemy attacks against coalition forces. We were tasked to support elements of the 1st Infantry Division's 3rd Brigade Combat Team in the conduct of civil military operations throughout the Diyala Province. Civil affairs had a unique dichotomy - one morning we might be building relationships with the locals, and that afternoon, we might assume a de-facto soldiering role on a kinetic patrol with the unit we supported. It forced you to be adaptable to the environment and mission.

We had maybe 50 Americans in the compound, including Coalition Provisional Authority (CPA) personnel, and their security detail comprised of Triple Canopy contractors, former tip-of-the-spear special operations veterans. Every day around 6:00 a.m., you would hear the sound of IEDs (improvised explosive devices) - roadside bombs - being detonated in the surrounding area by the route clearance teams. Baqubah was notoriously nicknamed, "Baqaboom."

Our first vehicle was a very primitive HMMWV with a short flatbed. I manned the M249 SAW on our early missions, standing up fully exposed. However, since our team Sergeant didn't like the idea of me bouncing around back there, I brought bungees with me and rigged myself to the pintle. He insisted that wasn't going to make a difference; he was right. HMMWV rollovers were a big cause of casualties. But the job had to get done! It was always interesting to see high-ranking officials on the large bases driving armored vehicles to go back and forth to the chow hall - while we teams in remote, dangerous locations, running missions on dangerous roads had the basic vehicles. Eventually our team was granted an upgrade and made a trip to Tikrit HQ to pick up our deluxe, shiny new M1149, complete with four doors, armor, and a spinning turret!

We ran frequent missions to the "Blue Dome" government building down the street so our leadership could meet with Iraqi leaders. The meetings were often held at the same time, which created a pattern of our movement. The trip wasn't long, and the drivers would drive like hell down the middle of the street over the island barrier through the gate and into the Blue Dome parking lot where they would combat park for quick egress. I was up on the gun - that is, I manned the M249 SAW from the turret. We'd lock and load and go down the long throughway, fortified by T-walls and rocket propelled grenade (RPG) netting onto the Iraqi street. Sometimes the CPA lead would attend the same meeting as our leaders. I liked this because that meant more armed friendlies on the trip and at the Blue Dome, in the event things went south, since the Triple Canopy contractors would drive them in their armored civilian cars.

From the minute they arrived at the compound, I watched how the Triple Canopy contractors worked, moved and employed specific tactics. I tagged along and asked questions as they did their security assessments of the compound. Many of them were seasoned retired operators so I appreciated the chance to watch and learn from the best and from years of experience - literally a whole team of professionals. They let me come and train on the range with them and their vehicles at FOB Gabe. I sat in on the mission brief and observed "what right looks like". Everything had a place; everything had a purpose. Details mattered. Fundamentals were everything. Communication. I appreciated the trust they gave me to do live fire training with them, but I know I carried myself as a Soldier who wanted to be there and become even better. I have carried over what I observed and learned from them throughout my life since.

When we weren't running missions to the Blue Dome, a project site, or another base, we met with local Iraqis in the small building by the south gate. Each member of our team was tasked to work with Iraqi leaders and the CPA on different focus areas. I was assigned the oil and communications sectors to coordinate and assist with US-funded and supported projects. Not a lot got done with the oil - imagine that - but I was successful with obtaining funding for infra-structure projects that benefited the Iraqis, including an installation of an emergency communications system for remote villages, repair of a microwave station facility, and the establishment of an internet café. One day my contact in the communications sector gifted me a cool book of Iraq postage stamps as a thank-you. That felt pretty cool to know I was making a difference despite the ongoing war.

Local Iraqis would come to our compound if they sought solatia payments (compensation for grief or loss) for the loss of their live-stock during a US raid or if they had information to share or needed help from the US forces with something. We used a team of local interpreters (terps) to translate for us; one of my duties was to manage the team. They made my task simple as they were very diligent and professional. We were lucky to have such dedicated interpreters who risked their lives every day coming to work with the US forces.

We wore many hats; Joyce and I also had to physically search any Iraqi women visitors. We would do this inside a small concrete bunker by the gate so if they had explosives on them, the explosion would be contained. I definitely wasn't crazy about that detail, but thankfully we didn't have any issues.

Since we were in the middle of the city of Baqubah, US forces patrolling through the area would occasionally stop by our small com-pound. One night as I was drifting off to sleep in a room next to the TOC (tactical operations center), I dreamed that I was right smack in the middle of the scene in *Saving Private Ryan* with enemy tanks rolling through. I could see the turret turning toward the building I was in, and I awoke from a very vivid dream gasping, "Sticky bombs!

Sticky bombs!" trying to disable the tanks like Tom Hanks' character did in the movie. My heart was beating fast, and I quickly realized we had visitors right on the other side of the wall where I slept - M1 Abrams tanks. The irony of the sounds of them moving in to the compound late at night had permeated my sleep and subconsciously connected them with a scene from a WWII movie – how funny! Wide awake now, I went outside to watch these amazing massive beasts in our small parking area.

One of the coolest perks about being deployed in a combat zone was a program that the Army offered that allowed you to put away up to $10,000 of your pay while deployed and guaranteed a 10% interest rate. I was all over that one - no-brainer! Growing my money while getting paid to do the coolest job in the world... win-win!

The Alamo

When your first major firefight is a siege....

Many people are familiar with the series of events in Fallujah in March 2004 when four Blackwater contractors were ambushed, killed and hung from a bridge. That led to a massive US-led assault to recapture the city eight months later, the first Battle of Fallujah. Complex, coordinated attacks against several US compounds throughout Iraq ensued immediately after the deadly ambush.

During the week of April 6, 2004, we watched from the rooftops of our building as angry crowds of hundreds of Iraqi protestors gathered in the streets around our compound chanting "Go Home, America" and other anti-American slogans.

Watching protestor crowds chant and march past our compound in the middle of the city. 2004.

Intel was spot-on about the time of the attack and at the projected time our little compound was assaulted for several days. Many have seen the video footage of Blackwater contractors on the rooftop in a similar compound under assault in Najaf. We were going through the same thing in Baqubah at the same time, except with Triple Canopy contractors defending the CPA principles, and by proximity and extension, the rest of us. Accordingly, they took charge and directed us Soldiers on which sectors to cover. Nothing spells unity of force like when your compound is aggressively attacked.

The CMOC was in the heart of the city, about 5 km from the closet forward operating base - FOB Warhorse. Since the insurgents were trying to cut off road access to us and isolate us, we called the compound "The Alamo."

The city got disturbingly quiet. And then it started. The fighting came in waves. The insurgents, who were part of Muqtada al-Sadr's Mahdi Army and wore black shirts and pants, would fight hard, and in our case in Baqubah, they would come back afterward with wooden coffins for their casualties. The surviving fighters would vanish into surrounding streets and date palm groves and then they would come back. Our infantry Soldiers on the rooftop were dug in on sandbagged positions firing Mark 19 grenade launchers and machine guns. I watched the Triple Canopy contractors ducking out of cover to take well-placed shots at insurgents climbing buildings around us - racking up a high enemy body count.

Fighting was day and night, but mostly at night due to the intense desert heat. But we had night vision goggles and could see approaching insurgents.

Looking out for approaching insurgents from the eagle's nest on our rooftop. 2004.

All throughout, we were thinking of the oh-shit scenario: a helo may not even be able to land on the rooftop of our building with all the direct and indirect fire, if we needed a medevac or to evacuate the compound. Direct hits and airburst mortars designed for broader damage at times killed the lights in the buildings. We felt and heard the shake and rattle from each mortar and RPG hitting the hardened building - our sole solid cover.

No one knew the ultimate outcome, but make no mistake, the random funnies and humor that emerged in the lull between the fighting is one of the purest and most innate and sacred properties of sustained warfare. I could share something here, but without context the humor would be lost. You had to be there.

Enjoying post-siege cigars with the gallant Sgt Kevin Whelan. 2004.

Years later I watched the 2016 movie, *13 Hours: The Secret Soldiers of Benghazi* - the story of the September 11, 2012 siege on the US Consulate in Benghazi, Libya, where Islamic terrorists killed US Ambassador J. Christopher Stevens, Sean Smith, Tyrone Woods and Glen Doherty.

I was instantly surprised by how many of the scenes in the movie inadvertently and vividly captured exactly what the multi-day siege looked and sounded like from inside our compound. Literally - starting with the scene of how precise and accurate the intelligence projections were for a hard time for the start of the attack and the way the fighting came in waves. The GRS guys in the movie reminded me of the Triple Canopy team and what I watched them do firsthand. Grateful they were there with us.

It was eerie to "relive" our siege through a movie about another attack, years apart. Unlike Benghazi, our three-day siege resulted in no US casualties. Due to the leadership and skills of the Triple Canopy contractors and the all-hands fighting of every armed US personnel, the insurgents were not able to breach our perimeter. Unlike Benghazi, our siege was part of operations in a combat theater. As we know, Benghazi could have been prevented if Americans got the assets they needed. Rest in peace to the Americans killed in Benghazi.

Battle of Baqubah I - 2004

We received intel that enemy activity was getting heated in advance of the Transfer of Authority from the US-led CPA to the Iraqi interim government scheduled for July 1. On June 18, 2004, US troops near us engaged in a 12-hour firefight, killing thirteen insurgents. On June 24, Al Qaeda laid out ambushes for the local mechanized infantry units that led to intense fighting. We could hear the nearby sounds of battle nearby from our location - the orchestra of small arms fire, machine gunfire, RPGs, and the low bass of daisy-chained IEDs. Stray rounds from the fighting a short distance away hit our chow trailer, grazing one of the contractors. We heard the real-time transmissions when Captain Christopher Cash was killed in the firefight. It was intense.

I later learned of the heroic actions of the platoon Sergeant for the North Carolina Army National Guard, whose citation for the

Silver Star reads, "Sergeant First Class Stephens crossed 50 meters in open terrain while small arms impacted all around him as he ran. He mounted the turret and pulled his Soldier out of the hatch, then lowered him to the waiting medics as rounds impacted the vehicle and other soldiers drove to safety. He returned to his vehicle under continuing fire, reorganized the platoon and led the move to FOB Gabe, fighting on as his Bradley was hit by a rocket-propelled grenade and his gunner was severely wounded."[3]

On June 25 a QRF (quick reaction force) team was ambushed nearby at the soccer stadium, which was also being used by insurgents. A call for fire went out, so we watched from the rooftop as F-16 fighters dropped a JDAM on the stadium. The mushroom cloud was epic! You could feel the blast wave as it passed through the solid hardened building beneath our feet. The might of the US military in action is a powerful force!

A JDAM decisively ends problems.
The mushroom cloud, as seen from our rooftop. 2004.

3   "The Hall of Valor Project: Chad M. Stephens". *Military Times.* https://valor.militarytimes.com/hero/3877.

Two days later the fighting reached another level around us. You could hear the mosques inciting violence against the US forces. Though I did not understand what they said and we relied on interpreters, I remember the distinct tone of their rhetoric. The June heat was intense and reports stated that the majority of our guys fighting on the streets were getting IVs to keep going. Once again intel was spot on with the time of attack, and our building was rocketed by mortars and RPGs – as was the Iraqi police station across the street and the Blue Dome. The insurgents were trying everything to target their city's government before Iraq became sovereign again. One Soldier on the rooftop got the proverbial million-dollar wound on his backside, but we were otherwise casualty-free.

Even after the Transfer of Authority took place, Baqubah remained a hotbed of activity. On the fourth of July, two Syrian insurgents attempted to drive a VBIED (vehicle born IED) - a car bomb – into our compound. The quick-acting Iraqi Civil Defense Corps shot the driver before he could get in, and the car burst into flames not 50 meters from where our team was located that morning.

Smoke from the neutralized would-be VBIED. T-walls and RPG netting fortified our small compound. 2004.

I was promoted to Sergeant that summer. It was a defacto time-in-service promotion, not a meritorious-based promotion. But it was still cool that the ceremony was in our compound and not stateside.

War selfie. Jersey girl standing on a Jersey barrier. 2004.

On July 28 we were physically rocked as a VBIED - a mini-bus - detonated down the street, killing 68 Iraqis, many who were in line to join the Iraqi police force in Baqubah. The blast was massive.

I did not let myself get close to the interpreters and perhaps was a bit standoffish. They went home every night - and I guess I always thought that with the craziness out there, they may not come back - and I did not want to get too close and feel the pain of more loss. One day one of our interpreters, Wafaa, a beautiful young lady, was killed on her way to work in our compound. Another was shot. Things were getting heated in every way.

THE LETTER

Someone I loved once gave me a box full of darkness.
It took me years to understand that this too, was a gift.

–Mary Oliver

Everyone has some sort of struggle in their life. For some, it's a debilitating disease or health issue, for others it's emotional abuse or physical or sexual assault. For some it's racism or discrimination. For some it's overcoming a socioeconomic struggle. My biggest hurdles to overcome were betrayals by those who you think would have your back the most: spouse, family and military leadership.

My unit rotated back to the States early in October of 2004. Several of us were dismayed about this early return because we had been slated to join the operations developing in Fallujah - the Battle of Fallujah, which not only regained US control of the city but avenged the killing of the Blackwater contractors.

We had a week of outprocessing in Fort Bragg and then returned to our respective states. I was happy to be reunited with my motorcycle and put on my good ole' broken-in Doc Martens boots but couldn't get my feet in. I asked the Marine if he threw them in the dryer or something. Of course, he didn't. This didn't make sense. My toes were scrunched up, but they fit just fine a year ago. I guess the weight of wearing a full kit everyday further flattened my flat feet. Great.

Tensions had been high in my marriage in recent years, and being deployed - physically separated with limited communications - did not help things. So while I had returned home, we were in separate bedrooms.

It was a rough first tour, having lost my battle buddy a few days after we got to the desert and dealing with the guilt of why her and not me. The siege in June was a bit hairy, but everyone in the compound came back alive and intact.

The war's demand was forcing the Army to grow, and new units were being created to meet the demand. For example, the only active duty civil affairs unit had traditionally been comprised of former US Army Special Forces personnel - "Green Berets" - but eventually it opened up to all Soldiers, contingent on passing a selection test. I determined I loved soldiering and wanted to join an active duty unit in Fort Bragg, NC, home of Army special operations. The Marine was not keen on it but was open to relocating. My focus upon returning was to get in the best ruck marching shape as possible to maximize my chances of admission into the unit. I was determined to train and serve with the best - and nothing was going to stop me. I reserved a hotel room on the post for six weeks of training so that I could acclimate to the hilly rucking terrain and environment. Even in October in New Jersey, I noticed my body was a bit more sensitive to the early season's cold. I chalked this up to having just returned from months in the desert where temps reached as high as 129 degrees F.

On Halloween night, I had only been back home in New Jersey a week or two, and the Marine and I were fighting about something. He went to the garage, got on his motorcycle and said he was going to meet his friends. I knew he was going to the bars. I said, "Fine, go, but call me and I will pick you up because it's raining and I know you are going to be drinking." I went about whatever I was doing. The evening progressed and turned into late in the night. I was concerned because I knew the bars closed at 2:00 a.m. and he had not called. I got in the vehicle and drove by the bars I recalled he used to frequent but did not see his motorcycle.

The rain was steady. Feeling uneasy, I went back to the house and jumped on the computer. We shared passwords to our email accounts so I hopped on his to see if any recent exchange with his friends would indicate where he was or who might know. It dawned on me that I did not know who his friends were now. I had been gone almost a year and there were new neighbors in our little neighborhood. We didn't communicate much. I had one phone call with him on the Thuraya satellite phone while in Iraq. But I admit I did not want to speak with voices from home because I needed to focus on the mission and did not want distractions.

I scrolled through his inbox to no avail. However, I saw one subject line that caught my attention. I opened it up and saw it was a string of messages between the Marine and members of my family. There was an attachment with references to edits. Curious, I opened the attachment and read it. As the rain turned from constant to downpour at the wee hours of the morning, I read a three-page draft letter addressed to a state elected official, signed by the Marine. The letter was an attempt to diagnose me as bipolar in an attempt to dissuade the Army from allowing me to go on active duty! It was signed by the Marine, but I recognized it was written by my father, the engineer who decided he was a qualified doctor - a psychiatrist, no less?! WHAT.THE.F#@K.OVER? I was in denial over what I was reading on the computer screen.

There was no date on the letter. I had no idea if it was mailed. My immediate thought was Holy shit, what are they doing? If anyone in the Army saw this outrageous letter full of unfounded and outrageous allegations, I could not pursue an active duty career! What was going on? I felt - *hunted*. There is no other way to describe what I felt while reading this, realizing my controlling dad had spearheaded some crazy plot to intervene in my life. Seriously, WHAT.THE.F#@K.

OVER??? I was exhausted. Through the chaos, I thought clearly, Let me deal with one crisis at a time. Plus, there was a good chance this was all a plot in draft mode and the letter was never sent. I would deal with this later. Right now my spouse could be dead on the side of a road from a drunken motorcycle crash. I did not want to deal with more death this year.

I called my mom, who lived nearby and asked if she had heard anything from the Marine. She said she had not, and I asked her to call me if she did. I said I was thinking of calling the police. I hung up and passed out, not knowing what to do, exhausted and temporarily defeated from the long, dramatic night. I woke up to the phone ringing a couple hours later. It was early, maybe 6:00 a.m. When I realized it was Mom, I asked, "Did you hear something??? Should I call the police?"

She merely said, "Why don't you come over, have some coffee?"

"Huh?" I thought and replied, "Coffee? The Marine might be dead on the side of the road. You *never* call and ask me to come over for coffee! What is going on?"

I feared what she would say next. I feared she had news and that he was dead.

"Hold on. The Marine is here."

I breathed a sigh of relief. He was alive! "Did he just get there? Thank goodness!" I replied.

She said, "No. He was here all night. He told me not to say anything to you because he wanted you to suffer, worrying about him like he did when you were in Iraq."

WHAT. THE. F#@CK. Are you freakin' kidding me?

The Marine had been located. When he came home, I asked him about the status of the letter. He had indeed mailed it. I told him to get the original back. Unreal! I packed my bags and set out to pursue my plan of going to North Carolina to train and prep.

First I stopped to vote on Election Day in the general election between John Kerry and incumbent President George W. Bush. The hits kept coming, as the election officials said they had mailed my absentee ballot to me overseas. I explained that our unit returned early, so I never received it. They still did not allow me to vote. I was devastated. As I walked outside, I saw the Marine coming in to the building to vote. I learned we had opposing political views, a characteristic I hadn't considered when I fell in love with him at 20 years old. Heck, at that time I had zero interest in, or knowledge of politics. I told the Marine what had happened and he laughed. This made me loathe him even more. I left the polling station feeling very defeated. Fortunately, President Bush won the re-election, even without my vote.

The office of the elected official whom the letter was written to eventually contacted the Marine and said that the allegations he put forth in the letter were quite serious, and asked if he really intended for them to look into them. He declined, and requested the original letter be sent back. It was returned in an Official Business envelope from the elected official. I retrieved the three-page web of lies and put it away.

RESILIENCE

I put the letter away and did not read it for 15 years, until I re-read it for consideration for this book. I was angry for a moment, remembering the kind of control that was directed toward me at that time in my life - at 31 years old, no less! Then I reflected on how free my life has become because I made the decision to find my freedom despite not knowing how I would do it and having to figure it out. The point is, no matter what level of WTF moment you go through in life, you have the ability to make a choice: let life control you or seek more control of your life. It may not happen in a day, a month or a year. It is okay to lick your wounds, heal a little, regroup, assess the situation and drive forth. But don't dwell on it!! One thing is for certain: if you choose to be the victim mentality, you will never be free of it. Life is full of battles you don't see coming, and you need to focus on surging forth and dominating the situation when things get crazy again.

Epic fail

It was an unusually cold winter spell in Fort Bragg, NC. There was no wind but the temperature was 7 degrees F. In the dark of the wee hours of the night, I struggled to push through the individual task at hand, losing feeling in my fingers and toes, feeling numb and dehydrated from the effort. I kept pushing forth with what I had to do, one step at a time. I had made it this far with my never-quit, can-do-anything attitude. I could not quench my thirst as the the water in my CamelBak hydration pack was completely frozen. Channeling my mental toughness, I completed the task, but my time did not cut it. The immediate feeling of failure was a force to reckon with. I was only minutes from the standard, but unequivocally a no-go. Weeks of disciplined training did not cut it. I failed.

I stayed with some Army friends in Fort Bragg while trying to figure out what to do. I had entered this endeavor with the mindset that failure was not an option. Because I believed I would win, I had no backup plan. In one moment I was dismissed for failing to negotiate a standard. My bad. My fail. But what now? The emptiness of not knowing the next step was as cold and unforgiving as the winter temperatures that week. I looked at my options. What did I have waiting for me at home in New Jersey? A broken marriage. A not-so-warm or -welcoming family. A city that almost seemed to have forgotten that which united us only three years prior during the darkest day of 21st century America. The thought of going back to work in information technology in corporate America did not even fit in the picture after ten months in Iraq. I was broken-down, confused and at a loss for my future.

Though my friends would have let me stay with them while I figured things out, I knew sitting around without a plan was not going to work. I needed to regroup and take action. My share of finances was depleting from living at the Carolina Inn on base, so my first initiative was to stop the financial drain. I grabbed the little bit of stuff I had, packed my vehicle and drove back to NJ to the home I shared with the Marine. I contacted my reserve unit in Staten Island and asked to be put on the next rotation, preferably to Iraq because although Afghanistan was an option, Iraq was where the action seemed to be.

In the interim while I waited for the next rotation, I decided to use my tuition assistance benefits from the military and enrolled in graduate school online so I could at least be working on something useful and productive. While the thought of getting an MBA was once a priority in my life, studying asymmetric warfare was much more relevant and appealing to me now. I was very grateful that one of my professors, Dr. Steve Greer, was a US Army Special Forces Senior NCO, who also served on the Office of the Secretary of Defense Military Analyst group. His courses were so relevant and fascinating. I studied and enjoyed learning about generation warfare, empirical history, global insurgencies, and guerrilla warfare.

ROLL THE THROTTLE

Do you want to know my secret to getting past a problem? It's really quite simple: Accelerate and move past the problem as quickly as possible. How so, you ask? Add more problems to it. Throwback to the self-reliance section! If you are focused on one issue and another issue comes along, it helps you reprioritize the real issue. It forces perspective and makes you think more clearly about what is going on. The less time you have, the more you focus. Think about it.

We have a natural tendency to delay decision-making and taking action because we fear the inevitable fallout or pain of an unsavory decision or action. The sooner you cross the line, the sooner you can move past the problem. Roll the throttle.

Think about times in your life when something happened, and you thought that things could not get worse. Then, things did get worse because something additional happened, creating a new problem for you. These are the moments in life that truly define you. The good news is: we each have the resourcefulness to handle a lot more than we think. If it hasn't happened to you yet, know that it will. Life will test you. And you have what it takes to emerge as the victor - stronger and more resilient.

A funny thing happens when your problems start adding up: some of them merely go away. How so, you ask? Here is the thing - having a problem is as much a state of mind as it is a literal resource sucker. Let's not ignore the fact that many of us tend to add much more to a problem than needed. We do this by speculating how it will get worse or by considering every bad imaginable outcome related to the problem. So when faced with five major issues to solve, a clear winner will emerge, and the weak link will subside in context of everything else you face. State of mind. Breathe. Assess. As sure as the sun rises and sets each day, you will push through if you think clearly.

Granted, some clear, nonnegotiable problems will automatically rise to the top. If you cannot breathe and obtain oxygen, that's a problem. If your body stops producing nitric oxide, that's a problem. If you cannot stop the bleeding, that's a problem. I think you get where I'm going with this. There are problems, and then there are *problems*.

In combat, friends are killed but the mission drives on. It has to. When problem after problem piles on and your bandwidth is stretched across multiple problems, it forces you to sort and prioritize so you can allocate resources toward the problems intelligently. The Pareto rule states that 80% of your problems come from 20% of your problem areas. Identify those problem areas - whatever or whomever they are. A problem area could be an employee in your business or a person in your inner circle. Don't let little problems become bigger problems because they are the only problem on the plate. Use context and build that toughness to handle conflict. Life can always be worse.

When you have multiple problems on your plate, it is likely one may resolve itself. This gives you a psychological win in "you vs. life". If you only have one problem, how miserable are the periods of time when you are in "wait mode"? If your plate is full of problems, you won't have to deal with waiting as you will be preoccupied.

I totally joke now about my first-world problems – such as a particularly humid day that will weaken the glue on my fake lashes. But there are people who that would cripple; they would literally shut down over something so petty. Do you see what I mean? Build that threshold in life. Your bad day may be someone else's easiest day. I am grateful for the problems I've conquered in my life because as much as they sucked, they've made me stronger, resilient and more capable as a human being.

SEND ME AGAIN

Life is 10% what happens to you and 90% how you react to it.

–Charles R. Swindoll

I enjoyed my studies and had completed 50% of the master's program in six months before an opportunity to deploy again came up with another subordinate unit. The unit administrator at my reserve unit ribbed me saying, "Khalifa, why in heck would you go overseas with another unit? They will never take care of you like we would." I didn't understand what he was saying. Take care of me in what way? I was a Soldier; I would be armed, fed and paid as such, regardless of the unit I deployed with. But he was a wise unit administrator and he understood the system and how it worked. He foreshadowed what I would experience. Little did I know.

I went on active orders late in 2005 and went back to Iraq in April 2006.

When one of my fellow Soldiers found out I volunteered to deploy again, he said, "Khalifa loves war," which sounds crazy - but I did. I loved the sights and sounds of the battlefield, our US military might in action. *Of course* losing fellow Soldiers was absolutely horrible and I hated to see any kind of human suffering. What I loved was the kinetic side of war - the distinct sounds of belt-fed badness, watching and hearing Navy jets taking off on the runway at the LSA, off to drop a motherload of hate and discontent on our enemy. I loved the whirr sound of the M1 Abrams, the sounds of Apache helicopters seemingly coming out of nowhere and flying overhead as you cruised down the MSRs (main supply routes).

Perhaps hearing and seeing the military apparatus "live" brought me back to the safe and cozy place in our living room, with Dad in the recliner and the war documentaries on the television, with the volume so loud that Mom would yell at him from the kitchen to lower it. But you can't lower the volume on the sounds of freedom and America's might!!

I loved soldiering and actively participating in the fight. I loved the fellowship of common purpose and serving alongside other motivated Americans. I loved rolling out as the turret gunner, especially when in the lead vehicle. It was a total crapshoot, but whether or not we were inching along to avoid driving over pressure plates, or flying down the middle of the road, there was forward momentum; there was movement. In my mind if I was going to be directly hit, it would be instant and there would not be much that I could do, so why stress over what I couldn't control? It really was not unlike riding my motorcycle on the highway, enjoying the moment as I scanned for threats.

At this point between Operation Iraqi Freedom and Operation Enduring Freedom in Afghanistan, the civil affairs force was a hot mess, piecemealed with Soldiers from different units and now Soldiers from the Individual Ready Reserve (IRR). Teams were mixed with students who chose this path and were "schoolhouse-trained" by SWCS at Fort Bragg, and those who were involuntarily reclassified into the job by the Army. These Soldiers did not exactly sign up for the mission, and many did not want to do it. Of course, they did not have a choice. They attended a mini civil affairs course that did not have the special operations requirements and standards that schoolhouse-trained students were required to meet. The IRR was the pool of Soldiers who were at the end of their contractual obligation, and therefore not actively drilling each month, but still subject to recall.

The team I fell into during our train-up period was led by an officer from the IRR. I met some great soldiers in my new unit with a great attitude. One of my best friends in this unit, Sean Cummings, was a first responder during 9/11 as part of FDNY's Rescue One. We enjoyed working out together in Fort Bragg outside of our required training.

I was in peak shape, and it felt really good keeping up with the pack in our small group of motivated Soldiers, as we ran through the woods. We trained together, drank together, and bonded. Incidentally, I was the only one in the greater unit that had deployed to Iraq before. It felt good to be able to offer relevant leadership and teach TTPs (tactics, techniques and procedures) that we used, to fight and survive in the desert. I felt like a valued member of the unit and was eager to deploy with them.

Team formation was like a lineup. I was very happy with the team I was placed on. The officer, "Roger" from the IRR was a chill, fun cat, easy going and happy. The two Soldiers "Ted" and "Wally" were physically fit and motivated with sound fundamental skills in garrison (stateside). Let's do this! I was initially told I would be the team sergeant because I was the only one with prior deployment experience, but ultimately they made the decision Ted would be the team sergeant because he outranked me by a couple months. I couldn't care less because I did not join the military for ranks and titles - I just wanted to rock 'n' roll with my team and kick ass together. It seemed like a healthy team overall. Promising.

Carrying the load

Since I had last been there, Iraq had gotten worse and enemy TTPs evolved and were deadlier. Pressure-detonated IEDs were common and explosively formed projectiles (EFPs), which would penetrate the thick layers of armor HMMWVs we were now equipped with, killing everyone inside, were emerging. Our unit arrived in country and the teams dispersed to their respective area of operations. I watched as the gravity and seriousness of what we were doing set in to my "green" teammates. I literally watched fear take over and paralyze competence. Roger shut down. His attitude changed and he revealed he was bitter that he had been called up to duty a month before he was set to graduate from his MBA. He would sit in our vehicle on missions in the track commander (TC) seat and take a nap instead of listening to the important radio transmissions. He would tell us, "What are they going to do to me? I'm already in Iraq!"

Roger said he had a female superior when he was previously on active duty that he despised. "Okay, and...?" I said. Ted admitted he had a fear of being around a lot of people. I said, "Perhaps you may not want to be on a CAT-A team. I'm sure if you talked to 1SG and shared that he would pull you to a job at headquarters. You do realize we will be around a lot of people. A lot of people that don't speak the language and probably want to kill us." He just shrugged.

On separate occasions, Wally said to me, "Shouldn't you be home with your husband, having babies?" Another time Ted said, "Why would a pretty girl like you volunteer to serve another combat tour?" and, "Why are you not a mother yet?", and "Why can't I just lock you in your room?" Seriously, dude? It was like aliens invaded the bodies of these Soldiers. They were not the same people I met in Fort Bragg. People change, and what better than war to bring out their true colors? I figured they were just scared, so I brushed it off and felt compelled to step up my game and keep them safe. It was a weird dynamic, but the protector in me wanted to simply keep them safe, as long as we were stuck together as a team.

Some of the more fluid missions were when we rolled out with our local Special Forces ODA and PSYOP team – each element supporting each other, such as in a MEDCAP (a medical civil action program). While the ODA team's medics were treating the locals and whatnot, we would help the PSYOP guys pass out their leaflets.

One time while on a mission with the Iraqi Army (IA), the IA in the vehicle behind me kept pointing his crew-served weapon toward the front, right at me. I couldn't leave my position, so I asked Roger to do something and he said, "Just duck." What an ass.

Missions were sometimes long. Obviously I dared not leave my position behind the M240B on top of the HMMWV because you never knew if a sniper was out there in the palm groves watching, waiting for you to become a soft target. To pee without leaving the turret, I would cut the top off the liter-sized plastic bottles we drank water in and drop the bottle down my uniform pants and pee into it standing up, all the time facing my sector so any enemy watching never saw me move away from my weapon. So there I was in the desert, multitasking as my teammates opted for naptime instead of readiness, zzzz's instead of scanning their sectors and listening to the radio comms - Yeh, bro, I've got this.

It was hot as hell, so you needed to keep drinking water and stay hydrated. One extremely hot day I spent four hours stationary in 120 degrees F heat, manning the gun in the turret. When we returned to base, I asked Ted if he would give me an IV. He said, no. I could not believe it. He was my team Sergeant, and Wally was on leave. We had all gone to combat lifesaver training together; he was qualified. Thanks for looking out, buddy! Did Ted check up on my status later? No. This is yet another indicator of what kind of "leader" he was and how he took care of his soldiers. I was just thankful that I wasn't critically injured and really needed his help, God forbid, in the middle of a mission.

Civil affairs is probably one of the most diverse jobs in the military, with a mission that finds you attached to a broad range of units, doing a broad range of things. That MP5 on the cover photo was not my standard issue weapon, nor did I have it for long, but it does explain the big smile! I laugh at the photo because you can clearly see the pen in the kidney pouch on my vest - because civil affairs Soldiers are always taking notes, with all the people we encounter throughout the area of operations. The flip side of the job is that we have lost a lot of civil affairs Soldiers because of the high rate of exposure to dangerous or vulnerable scenarios.

Civil affairs teams are the ultimate hitchhikers, and when we do not have enough personnel for an organic convoy, we rely on a variety of transports to get around, from air to ground. Loved when we flew around in Blackhawks! At one point our team was in an area where enemy TTPs resulted in a mandate for vehicles to drive very, very slowly to avoid EFPs on the MSR. A 15-mile trip took over an hour.

Several times we transported in the back of a Bradley Fighting Vehicle. Once the back ramp was closed, it was dark and extremely hot. We would stick a cold bottle of water between our body armor and body to keep cool. I joked that it was the ultimate detox - the Bradley weight loss program, where you would lose five pounds of water weight, in just one mission! And wait, there was more! If you acted now, they would throw in stinky enemy prisoners in the seats right next to you, to add to your VIP experience in the Bradley's combat sauna! Much respect to the badass crews of the Bradley Fighting Vehicles! These guys rocked, and I had a lot of great memories rolling with them.

Whipped by the chai

The team of terps I managed in 2004 were key to mission success and worked right beside us translating Arabic from the elected officials and Iraqi civilians we were meeting with. They supported us at provincial meetings and in ad-hoc situations – such as when an Iraqi came to a US base with a request or information, or on missions to villages where the locals did not speak English, as was often the case. As in any sustained relationship, trust is built over time. There are countless stories of Iraqi terps, who were integral and armed parts of teams, fighting gallantly along their US partners. At our CAT-A level, we employed uncleared, local terps. They were vital to the mission, and they all risked their lives working with US coalition forces.

Interpreter translating for local Iraqis and myself. 2007.

We received a replacement terp on our base named Huga. She was well-dressed, and in her late 30s. Her English was very good; she was Iraqi but allegedly had lived in England for some time.

The other terps didn't like working with her because they didn't trust her. She was from outside of the area, they didn't know her, and something didn't sit right. My "spidey" senses were on as well. I watched her and noted her dual personality. She was very quiet and nondescript around most of the Soldiers, but when she was just around our team – particularly around Roger and Ted – she turned up the charm, started making chai tea for them, flirted, etc. Wally was indifferent, but Roger and Ted loved the attention and how she would make them feel.

Huga knew what she was doing. I sensed there was a hidden agenda beyond a paycheck, so I continued to watch her like a hawk and maintained close communications with Soprano - the badass terp that worked with us on large-scale operations and had proven himself time and time again. We would compare notes on our observations, and both agreed something didn't add up.

As time went on, I observed minor things Huga did. I reported this to Roger and Ted, but they dismissed them. Of course they did! Ted even laughed and said, "The Iraqis would love a catfight!" Wow, dude. Huga was not a fool; she was very keen on the dynamic in our team and played on it, winning the hearts and minds of my team's leadership. I did not know what she was brewing besides the chai tea, but it was palpable, and it was not good.

One day we went on a mission to a big council meeting. In the room were 35 Iraqi leaders, all male, many older; Roger; me; and both Soprano and Huga. Midway through the meeting, Huga started a scene yelling at Soprano in Arabic. This was a total WTF moment, so out of place, and Roger and I, of course, didn't know what was being said, as it was in Arabic. Soprano had an annoyed look on his face but kept his cool. Huga got even louder. Picture a courthouse, and in middle of a hearing, the court stenographer starts a scene. That's what was going on here. So out of line.

Thinking quickly, I quietly put my hand on Huga's arm, turned my head to her and gave her a smile and whispered, "Huga, it's okay." She stopped screaming, shut up, closed her notepad, huffed and crossed her arms. Soprano had to finish the translating for the rest of the marathon meeting. Wow. That was weird. Did I mention there were 35 Iraqis in the room?

A meeting with local leaders. 2006.

Back at the base, Roger, Ted and I had a meeting about what happened. Soprano told us Huga was yelling about what one of the Iraqi leaders said. Wow. Not her place to get involved. She was there to translate, not to get involved. We all agreed Huga was a liability and her ruckus may have undermined our efforts.

Toxic leadership

The next day Huga came out of nowhere to me, hugged me, and said "How are you today?", very sweetly like we were the best of friends. I knew exactly what she was doing. She was panicking about her outburst. Maybe she even heard from the other terps she was being replaced. I had nothing to say to her. This was weird. She kept saying, "You are my best friend. You are the only one here I trust ...." The melodrama was ridiculous.

Roger walked in to the room, and said, "Ladiesssssss! Let's get to the bottom of this!"

My jaw dropped. WTF? It wasn't even as much the inappropriateness and lack of decorum of addressing a Soldier - an NCO - like that, but the fact that he was addressing us both **together**. I knew in an instant this was not going to go well.

Before Roger could speak about the incident at the meeting, like clockwork Huga turned on her Dr Jeckyll personality. She fabricated a lie that I barged into her trailer late at night demanding she translate papers. Wow… I couldn't even digest the WTF moment before Roger leaned forward, and his face literally changed. He pointed at me and said, "SSG Khalifa, this is a verbal counseling to you…"

He went on literally attempting to blame me for Huga's unforeseen outburst at the meeting, saying I was louder than she was. WTF was going on. It was almost like whole scene was orchestrated. I was waiting for the clown car to roll up on our base. This was seriously delusional.

I turned to Huga and said, "Huga, you know what you said is not true." She looked down. Was Roger seeing this? Here he was siding with Huga, believing the terp over his Soldier. And "counseling" me me in front of a terp. What has this world come to? I needed a witness. I said calmly, "Just one second, Sir."

There was **no way** I was going to let him talk to me like that with no witness, much less in front of a lying, manipulative terp who yesterday undermined our meeting's efforts. I ran outside and found a Sergeant. He came back in the room with me and suddenly Roger's tone changed. He no longer used the crazed, dramatic tone, but he did continue to spew some crazy things.

Roger continued, "SSG Khalifa, you're always trying to control the situation, take charge of things …". I was thinking, "Because you're a piece-of-shit leader and stuff has to get done."

Ted walked in the room and caught the tail end of this. The fact that we were having this conversation at all just blew my mind away! The fact that Roger - whether he believed me or not - even questioned me or my integrity - in front of a terp, no less - blew my mind away even more!

Huga was dismissed from the room and the Sergeant left. When it was just Roger, Ted, and me in there, I said to Roger, "I can't trust you. If we were on a mission, and fired my weapon and an interpreter said anything about me or what they saw, you're automatically going to believe him over me - your own US soldier who you work with every day." He didn't say anything.

Maybe my point was sinking in, so I continued. "Sir, from day one you have had blatantly open disrespect for me. You openly say you don't like women in the military. I cannot believe that just now you tried to counsel me in front of a terp! What are you thinking? There are rules when you counsel someone. Even Ted knows that." Roger nodded. He didn't disagree, nor apologize.

In the pause, something occurred to me: Could he possibly have planned this scene on purpose, in hopes that it gets him sent to Tikrit, for a cushy staff job? That had to be it. That was the only logical explanation. Unreal.

I then said, "Sir, just like you were hoping earlier that by disrespecting me with the women in the military comment I would do an EO (sexual harassment) complaint on you so you would get sent to Tikrit, that is not going to work now." He smiled. I knew it!

I continued, "I will, however, be telling our 1SG all about this because I am not going to allow this abuse you give me." He nodded. Unreal. Ted just stood there. The scene was pathetic. This was sad that an NCO, had to essentially "scold" him for what he did. And he nodded. He knows he did wrong. But does he care? Of course not. Roger has nothing to lose. In his mind anything that happens to him cannot be any worse than his job here as a team leader in Iraq. He always talks about how the minute he returns home and is released, the Army can kiss his ass.

Ted just stood there, looking down with his hands folded. The dynamics of this team are bizarre. I lived and worked in Bizarro World, like in *Seinfeld*. I turned to Ted and asked, "Ted, what do you think?" To his credit he said, "Sir, I stand behind SSG Khalifa in that what Huga did was wrong, and we do need a terp that we can count on."

Later on the Sergeant who I brought in as a witness said to me, "Roger is obviously intimidated by you." I said, "I can't believe I have to deal with this," and thanked him for being there.

So one would think that Huga was fired shortly afterward for the meeting incident. She was not. She continued to work with our team. She continued to do things that were suspicious, some of them direct security violations.

I documented and reported these to our 1SG who was far away on another base. I sent an email explaining the dynamics on my team to the Marine – even though we had started divorce proceedings. I sent it to him with a note saying if anything happened to me that didn't add up, he would have information and would know what was going on. He was a Marine. He knew this situation was unacceptable, and dangerous.

I kept going, one day at a time. I didn't share the internal team drama that was going on with the other Sergeants in the unit we were attached to because there was no way I could relay that level of crazy. And it was crazy. I appreciated every chance I had to go on missions with them. I enjoyed having a tactical mission and serving alongside and learning from active duty infantry badasses, but I also felt safest while going on patrols on MSR Tampa with them, than back at the base with my team. Do the math on that one.

Exercise helped me keep my sanity as well. Some days the roads were "black" (restricted, so no ground missions) and the internet was shut down as notifications of next of kin were made for recently killed Soldiers. Those long days sucked, as I was generally alone. I had grounded my teeth down pretty bad while I slept, internalizing all this stress of my team that didn't have my back and the shady terp whose intentions were still suspect. My canines were not pointy anymore.

Perception is reality: Fear

Sleep my friend and you will see
That dream is my reality
−Metallica, "Welcome Home (Sanitarium)"[4]

I had written down a dream I had during this time period. It was more of a nightmare. I'll spare you the finer detail. In the dream I was alone in our area of operations, trying to figure out how to survive and find my way back to coalition forces. There were blasts and a firefight during a mission we were on earlier on the day. Apparently, I was knocked unconscious and Roger made the call to leave me behind… Of course he did.

I had emailed the nightmare to one of my friends on another base. It was hard to read it all these years later because it's a reminder of the fear I was living in, on that team, with the shady terp. This was a warzone so it was not beyond the realm of possibility that a scenario like I imagined in my sleep could play out, given what was going on.

Obviously, I'm alive, well and writing this book thirteen years later. Unfortunately for many other Soldiers, their nightmares became realities in the desert. And I'm not even talking about firefights and IEDs. Many Soldiers - both genders - were victims of toxic leadership, sexual assault and other abuses that affected or haunted them for years afterwards.

---

4   Metallica, "Welcome Home (Sanitarium)," track #4 on *Master of Puppets*, Elektra, 1986, LP.

Enemy Exposed

Going on missions with Huga the crazy, shady terp was stressful. I did not trust her and did not know her true agenda. She continued to do things that compromised security, even while we were on the mission! I always reported it to Roger, and of course, he blew it off. I seemed to be the only one who cared. Three weeks of this went on. Not the best weeks of my deployment, for sure. Then one morning, right before a mission, we got the news: Huga was arrested for ties to Al Qaeda Iraq.

I'll never forget as we sat in the HMMWV. No one said a word. Then Ted glanced at me, looked away and broke the silence. "Khalifa, I hate to say it, but you were right." I didn't say a word. The silence was deafening. Wally next to me focused on his M4 between his knees. Roger didn't say a thing.

I didn't want to be right. Didn't we all want self-preservation and longevity for our team? Intuition is survival, and intuition proved its value. I was angry my own team did not trust their own fellow American Soldier over an Iraqi terp. They let their insecurities and personal disdain for me get in the way of clear thinking. And Roger even used this situation for his amusement, when he tried to mock me in front of what turned out to be our enemy.

No, nothing happened; no convoy was hit as a result of this, but who really knows what damage was done, what information Huga already collected and passed on? What did she really say or not say while translating for us at meetings? This did not have to happen. But they were whipped by the chai.

EMBRACE THE SUCK

Pain provides invaluable growth. Yes, it happened to you, but you can choose to react in a way that allows growth. Think about it. If you suffered in life and are still alive today, you probably learned a lot from it. The experience likely changed you. That is powerful. Maybe you changed habits or behaviors as a result. Or maybe it changed *you* and developed your best qualities.

If the pain was caused by people, it is a shame to let whomever caused it win even more by morphing *you* with a reduction of your happiness and inner peace. Don't let people steal your soul! They have already done enough.

In the immediate aftermath, you sometimes cannot make left or right of it, so shelve the situation and drive on with survival. It may take time, but when you finally have the headspace and timing to reflect upon it, look at the totality of everything. Work back to the root cause and gain a broad perspective. Understanding context can help you move on quicker.

Here is an example most people can relate to. Hunger. Are you really hungry and in need of nourishment? Or did you just smell something amazing cooking and want to experience the flavors from those aromas? Or are you actually dehydrated and thirsty and you believe food is the answer? Or are your cortisol and insulin levels off, and you believe you need food to feel right? Or are you an emotional eater and believe food will make you feel better? Or is your diet of processed food leaving your body nutritionally bankrupt? The point is, for just about anything when you accept the easy answer of "the obvious," you are doing yourself a disservice because you are not looking for the root cause. Knowing the root cause of something can lead to long-lasting and sustainable solutions. To find the root cause, start by asking yourself questions and answering truthfully. Use critical thought to dig deeper and find your answer. Upcoming chapters will reveal more "food for thought" on this example.

It took me a long time to figure it out but I eventually realized that Roger, and to a lesser extent, Ted, and Wally were all in shock after arriving at war. This was the root cause for their behavior. People do and say crazy things when they are scared. Remove them from the environment and they can bounce back to being normal. Human nature is rather interesting.

Silent, but deadly

Nothing to see where the sleeping souls lie
Chemical warfare

–Slayer, "Chemical Warfare"[5]

When we moved at 30 MPH in the cooler months, it made my eyes tear. I loved the rush of the wind, and missed my motorcycle! My cheeks would get warm and pink from windburn from being in the turret.

With SSG Sean Cummins and SGT Jeffrey A. Scotti
before a night operation. 2006.

---

5 Slayer, "Chemical Warfare," track #1 on Haunting the Chapel, Metal Blade, 1984, EP.

I was grateful to get to go on all these other missions with the unit we supported. I became much more mission capable working alongside Soldiers from different units, and they appreciated a willing NCO coming along. A few times I replaced a PFC or SPC on a mission and felt good about being able to do that, as I knew they probably enjoyed a random break.

Going down muddy canal roads after it rained was never fun. Though we moved slowly in the heavy vehicles, the threat of a rollover was a reality. Many Soldiers and Marines have died in vehicle roll-overs, some having drowned, trapped inside their Bradley Fighting Vehicles. Those reports were some of the grimmest.

One day Sgt. Vessy, Sgt. Jones, and PV2 Rice were hanging out in the shade of an Abrams in the parking lot, talking about the mission the night before that we were all on. PV2 Rice asked me why I was still wearing the combat patch of the previous unit I was attached to. "That's technically the only one I'm authorized to wear," I said. He ripped off his unit's patch off of his shoulder and replaced the one I had on. Coolest moment all year. That meant **a lot**, coming from one of the active duty privates.

Some of the missions were memorable because of the odd or funny things we saw or did. One time the infantry unit was trying to get inside a house without blowing up the door. One of the platoon Sergeants found a window that was small enough to crawl into. As I moved over to his position, I felt a kindred connection with the "tunnel rats" of Vietnam who were smaller men who could fit easily in the narrow tunnels, hahaha. Alas, as he hoisted me up, he got a call on the radio that they found another door that was open.

One day I was slated to join the explosive ordnance disposal (EOD) team on a mission to blow up 10,000 lbs. of stuff they needed to destroy. The convoy I was on was running late, so I missed the EOD mission by a half hour and was bummed. It was always a blast to blow stuff up! They weren't going far away, but I noticed that I never heard a large boom. Strange. About an hour later, I saw all the EOD guys back, and they had had a very concerned look on their face. I thought, Shit, someone got hurt – but no, everyone was there .... I didn't ask questions. Over the next week, our patrol base was locked down and there were a lot of visitors and unusual activity. I had a hunch it was tied to the EOD mission I had inadvertently missed.

We were eventually told what had happened on the mission. The mission was to blow up 40 arty (artillery) rounds that the Iraqis had turned in to them. When they looked inside one of them, they saw it was filled with a black liquid. Fortunately, the wind was blowing the other direction. They immediately aborted the mission, headed back, and called the chemical specialists...whose tests confirmed mustard gas! Bad stuff. WMD much?!

The rounds traced back to the Muthanna chemical plant. This certainly explained some of the strange missions we were a part of in recent weeks. I connected dots in my head, and also thought back to Huga.

Years later I read about the bunkers of sarin warheads UNSCOM found at the Muthanna plant after Gulf War in 1991, and the 500 missing mustard gas arty rounds.

Surge buildup

In early December 2006, Wally and I were transferred to FOB Warhorse to support the effort in advance of the Surge troop buildup. We grabbed the Chinook flight and arrived at Warhorse, where we joined different teams. My new team leader, CPT Darren Plotts, was very chill and just wanted to get us home alive. I was the team sergeant for our new team.

Reunited with others from our unit, I learned that racism and chauvinism ran amuck in our battalion. Wow. I suppose I was glad I wasn't the only one dealing with crazy personnel/personality issues on my previous team, but it was demoralizing to see how widespread it was.

The new team was awesome! We were attached to 1st CAV's 3BCT (3rd Brigade Combat Team). I really enjoyed going out on missions with them despite the volatility of the area.

Engaging with children while on a mission. 2007.

As Surge operations increased and fighting intensified, 1st CAV suffered many casualties. We attended memorial services for the fallen almost weekly. It was a dark time for all US forces in the area, with the high loss of lives.

One day our team received an unexpected surprise as we were awarded 1st Cavalry Division's Golden Spurs for supporting them on combat missions. I recalled the long lineage of the Division, and the fighting they did in the Vietnam War and felt so privileged to have been a part of their operations. I thought of the 1st Cav Soldiers we recently lost, and was humbled.

Drama, Inc.

My previous leadership – Roger and Ted - gave me my NCOER (evaluation) for the period of time on their team. You could not get a worse NCOER had you committed a crime. The entire evaluation was contradictory in itself – they didn't even make the effort to support what they wrote. They knew it too. They were smirking while trying to look serious.

In the portion of the seven Army values that spell the acronym LDRSHIP (Leadership, Duty, Respect, Selfless Service, Honor, Integrity, Personal Courage), the leader selects yes or no for each and provides supporting bullets. If an NCO gets a single no, that's usually a bad, bad Soldier. The only yes blocks they checked for me were Duty and Personal Courage – probably because I had the personal courage to stick with that crazy team. On top of it all, they did not have counseling statements to support their claims either.

Though I laughed at reading it, I was annoyed that they continued to screw with me – for sport, I was convinced. So I followed the process and appealed it. They had to rewrite it, as it was asinine. This was such a silly waste of time in the middle of a warzone. Alas, it didn't kill me, but it did underscore my desire to disassociate with negative, petty people. I rejoiced that at least I was back on a normal team and this drama too, would pass.

Christmas 2006 (Journal)

I woke up a couple hours before my alarm clock was set to go off. It's Christmas! I thought. I lay on the mattress-on-cot, under a warm hadji blanket. I smiled, remembering when I was a kid at home on Christmas morning. The aromas of the kitchen found their way to my room in the attic. Mom was always up early preparing to cook a turkey. Coffee brewed, and even though I did not drink it then, I still recall the percolating sound of the coffeemaker and the smell of the roast. I remember jumping out of bed to play with the toys Santa brought us late the night before.

I'd love to just call Mommy and Daddy from the "public" phone bank on the base, and wish them a Merry Christmas and assure them I am okay. I'd love to hear Mom's beautiful voice and accent. My parents are in their sixties and I wish I could be in their home - even just for the day - Christmas was always full of love in our home. Any one of the New York City radio stations plays 24 hours of Christmas music. Daddy had an RCA 45 record of the dogs barking "Jingle Bells." Mommy would probably laugh if I asked her to plug in that crazy electronic chirping bird ornament we would hang on the tree so I could hear its familiar chirp over the phone. But I knew I'd get teary-eyed and can NOT do that in front of Soldiers. Oh, no way.

This is my first Christmas in Iraq. Even though every day feels the same here in Iraq, this morning feels different. Maybe it's because I can roll around in my warm covers and stretch and just be lazy for a few minutes amidst the early morning calm of the FOB. I think of the family of a Soldier we recently lost. This is their worst Christmas ever. They are probably still in shock having just recently received the news and are now planning funeral arrangements. This Soldier was inside an Iraqi police station perimeter. The insurgents fired a mortar that hit him, a loader in an Abrams, as he stuck his head out of the hatch. Two others were injured, one very badly, from that mortar. F#@ king Buhriz. I know that police station. Classic shithole. I think about when we stopped at the Iraqi Army (IA) compound on Tuesday.

A new lieutenant had wanted to come on a mission with us, and conveniently backfilled as a driver because our team was short that day. He had stepped out of the vehicle to run over and take photos of the IA truck that was just hit by an IED as it staggered into the compound. I had to literally yell at him to STAY inside the vehicle. I jumped out, still connected to the headset and yelled at him to get back in. I would have honked first to get his attention but I couldn't reach over all the crap in the middle of the vehicle. We have the newest up-armored HMMWV - the M1151, but it has some design flaws to include the giant barrier between TC and driver sides because of the radios, Blue Force Trackers (BFT), etc.

I cannot help but keep thinking about this recent KIA from the tank company that supports us on missions. It just dawned on me that he was out there with us twice last week. He is probably in the photos we have from the big MEDCAP we did. Eerie.

Even though I miss my family, I couldn't have spent Christmas Eve with anyone better last night. I spent it with US Soldiers, defenders of freedom, gathered around the large fire pit in front of my buddy's trailer. We started off listening to Christmas music. Then Latina music, then trance music. One of the captains acquired some copper wire to burn. I stared in awe at the spectrum of colorful flames that emerged from the fire. Purple, turquoise, dark violet. It was beautiful. It didn't take much to amuse this "city" gal!

The captain, Greg Amira, was a World Trade Center survivor, who later founded the nonprofit, The Wounded Vets Association. On 9/11 he was working as a broker with Morgan Stanley, and was literally caught in the rubble. We listened to his story and saw the scars on his body. I had never talked to an actual survivor who was trapped under the debris.

One of the PFCs from one of the line units kept coming over by me but didn't say much. I knew what he wanted. He wanted to talk. And not to his buddies. I gainfully employed him by helping me assemble S'mores. His task was to hold the chocolate and graham crackers as I heated up the marshmallows on the end of a hanger over the fire.

Eventually we ran out of wood to burn and cigars to smoke. It was past midnight. I bade goodnight to the remaining bunch and retired to my trailer. Not five minutes later, I heard a knock. It was the young PFC. I asked him if he wanted hot chocolate, and he said yes and immediately sat down on my bed. He said, "Oh man, you have blankets!" and complained about the lack of warm blankets in his life here in Iraq. I made him the hot chocolate and sat down on a tough box. He admired my tricked-out M4 and red laser on my 9mm. I picked up the paratrooper's M4 with recently-added grenade launcher, and said, "Damn, that's heavier than mine with all the gear I mounted."

I asked him if he called home yet, or if he was going to call tomorrow. He hadn't called. He said he had no one to call. I asked him if he had children, having noted the wedding band. He said he had two, but he didn't think the girl was his. I nodded. I asked him how old he was, and he said, "Guess." I guessed 21 and was correct. So young.

The PFC spoke about a couple missions recently, and one when someone was killed. He stopped and looked at me. He said, "Sorry I talked your ear off. I just get tired of talking to the same guys all the time." I smiled and said, "I understand." He seemed finished, so I said that I needed to get some zzz's, even though I would have listened all night if he wanted. He didn't seem suicidal or anything like that. He got his weapon to go and stepped into the cold. He said, "Damn, I gotta walk to my pad in this cold!" I asked him if he needed a flashlight, but he didn't. I gave him a hug and said, "Merry Christmas." I'm going to the hadji mart today and going to buy him one of those warm blankets.

Murphy strikes

You never forget the mistakes you make in combat - even if no one got hurt. I forgot my night vision goggles on an early day mission. Mission creep took over, and we did not return until after dark. Thankfully we didn't have any issues on the route home. I always brought them with me regardless of the mission after that. Late in 2006 I was on a convoy that got lost en route from one base to another. Thankfully nothing bad happened to any of us.

It was a rare convoy for me in which I was a passenger being dropped off at the destination base, and not in the driver, gunner or TC position in our vehicle. I was, however, familiar with the route from having traveled it frequently a few months prior. I let the lieutenant in charge know this, and he placed me behind him in the same vehicle.

Murphy's Law states that what can go wrong, will go wrong. After passing a familiar checkpoint, things started to look... different, from my view in the back seat. I should have spoken up, but I trusted the lieutenant. I assumed he was following the Blue Force Tracker (a GPS system) and knew something I didn't. I'm not sure what happened up front but we got lost. Thankfully I had my personal GPS device on me with waypoints stored that helped us plan a route out.

Though the ultimate responsibility laid with the lieutenant as the convoy commander, I shared the burden because I had not spoken up when he made the wrong turn. This mistake, however, wore deep upon my soul.

ADAPTABILITY

You will make mistakes in life, but continue to be resilient and adaptive. Doing so will allow you to thrive. To do this you need to move on after making a mistake. We are flawed as humans, we make mistakes, so don't torture yourself. It is important to forgive yourself, or you will stop living. Is anyone perfect? No. All humans make mistakes, and if you wait for angels to come down and give you forgiveness, you risk never getting that forgiveness. Resilience requires forgiveness. I'm not talking about forgiving enemies and evil. I mean forgiving yourself for mistakes and moving on.

A prisoner who committed a crime serves a sentence, then he is released. The idea is that after paying for their crime, he has been absolved and can move on. Apply the same to yourself because likely the mistake you made would not garner a life sentence. We owe it to ourselves to live the best life we can while here on this Earth.

To forgive yourself, assess what took place that you have so much guilt over. First of all, what was the intent? Did you intend to cause harm or was it an accident? Did you accept ownership for the mistake? And probably most important: Did you learn from it? Are you better informed or equipped to avoid the mistake again? Every single great leader has made mistakes. This is a fact. And I guarantee some of the mistakes they made are far worse than something you did. If everyone who ever made a mistake condemned themselves to a self-imposed prison, humanity would not grow and civilization wouldn't progress.

Wallowing in guilt will not change what happened, nor will it improve anything. The best reaction is to take action. You can do this in several steps:

- Be real with yourself and take ownership.
- Face consequences head on.
- Develop a game plan to apply what you learned from a mistake like this.
- Teach others. A learned mistake is a powerful example.
- Make amends if it is necessary, forgive yourself, and move on.

Nothing changes for people who make mistake after mistake but do not take ownership or they take ownership but show no remorse. These people do not experience growth as a result of their mistake.

Battle of Baqubah II - 2007

Baqubah was getting more and more intense with approximately 2,500 enemy fighters with the ISI – the Islamic State of Iraq – claiming it as its capital. According to Dr. Bill Reeder, the chief of leader development at the I Corps Battle Command Training Center,

"Baqubah was arguably the most significant campaign of the war in Iraq after the initial invasion was completed."[6]

It occurred to me how sometimes in life you pay the price up front and reap the rewards later. Had I been team sergeant for the previous team instead of Ted, I would have been held back at the other base with Roger, and missed the chance to serve alongside a very elite team of warriors.

Our team was now attached to Charlie Troop of 5th Squadron, 73rd Airborne Cavalry Regiment, 3rd Brigade Combat Team - the first reconnaissance, surveillance and acquisition targeting squadron activated in the 82nd Airborne Division, otherwise known as Task Force 300. It was an honor to work with a unit that was awarded a Presidential Unit Citation for counterinsurgency operations in the Diyala Province.

Everything falls on leadership: the speed of the leader is the speed of the team, and leadership in this unit was tight and effective. It was awesome to see day after day what right looks. The level of competency and professionalism was next level. I learned so much from being a part of their operations. The noncommissioned officers were some of the finest, hardest charging warriors, a high caliber. Everything from how they conducted pre-mission briefs and after-action reviews to how they tactically moved and communicated on patrols was impressive. It's not that they were doing unprecedented things; it was more that they were doing soldiering fundamentals so well.

"Train like you fight" is a military maxim, but it is not always adhered to, depending on the leadership of a unit. It was clear that leadership had set the culture and climate for the unit to operate so cohesively. We had set up a patrol base in As Sadah in an abandoned schoolhouse where Al Qaeda had previously ran operations. It was nicknamed FOB Coomer after the kick-ass 1st Sergeant, John

---

6   Kramer, Don. "Lessons of Baqubah Now on DVD." www.army.mil, June 13, 2008. https://www.army.mil/article/9949/lessons_of_baqubah_now_on_dvd.

Coomer. Our team went on mounted and dismounted patrols with 5-73 CAV throughout the area. Going on foot patrols in the palm groves during the day and at night was pretty cool.

Mud walls and palm groves. 2007.

Our team was very well integrated with the unit. Due to the Surge's demands we split up further and I found myself in a team leader position. One day we went on a large-scale day mission in the "city center". It was a mission designed to show a peaceful US presence after some intense fighting recently. It was eerily quiet as dozens of Soldiers spread out. It was one of those missions that stood out in my mind, when you have an uneasy feeling knowing things could go south very quickly.

A voice started speaking from the local mosque's intercom as the tops of date palm trees swayed. We passed out Iraqi blankets and soccer balls to the people in the streets, as well as flyers that the PSYOP team developed and printed, with a coalition forces message.

Kids came from out of everywhere. They enjoyed playing with the soccer balls and when I whipped out a camera they hammed it up, singing and dancing.

The boy on the right is holding a PSYOP leaflet. 2007.

There were no issues on that mission. Later on, back at the schoolhouse I walked over to the burn pile in the back to throw out some MRE trash. PFC Daniels was burning something, so we started chatting about the weird things we saw while we were all out in town. He looked over at my shoulder and said, "You need one of these. You're one of us." He ripped off his patch and gave it to me. It was one of those moments that humble you.

I always felt a bit of imposter syndrome because civil affairs was an airborne unit and I hadn't been to Airborne school yet, as the timing hadn't worked out. A Leg is a Soldier who hasn't been to Airborne school. And now this PFC was giving me the 82nd Airborne Division's patch – *his* patch, even. I accepted it, and smiled, saying,

"You do realize you're giving this to a Leg, now?" - to which we both laughed.

I was honored to have the chance to roll with these warriors and was always watching, observing and learning from some of the Army's best NCOs, mission after mission, and in between when we lived at the patrol base. One of these badasses was the medic, SFC "Doc" Benjamin Sebban. It was seriously a privilege to have walked among so many heroes, many who are still serving today.

Standing next to our vehicle. The lethal truck bomb exploded right behind where I was four days later. 2007.

The compound was tiny. We parked in the corner of the compound, alongside the PSYOP team and their vehicle. We all slept in our vehicles at night. I remember being wrapped up in my Iraqi blanket, laying in the turret next to my M240B. I looked up at the dark sky, and as my eyes adjusted, I could see the outline of a Raven, the unmanned aircraft system flying overhead. I was glad for the overwatch, as it was transmitting live video back to the tactical operations center (TOC). On March 16 we rolled back to FOB Warhorse to refit. The next day we felt and heard a thunderous boom.

A truck bomb had plowed into the patrol base, detonating when it got stuck in the mud. Doc Sebban was the first to see the truck bomb and screamed for everyone to get down. He sustained wounds along with dozens of others, but bled out before the medevac arrived. The news of his death was absolutely brutal; Doc Sebban was very respected and liked.

Our vehicles had been staged for several days in the corner, right in the kill zone, meters from where the truck entered. If the PSYOP team or our team were still there, we would have surely had casualties due to proximity of the blast. Yet fate had us leave the day before. Our angels were working overtime.

In an instant

On March 25, our convoy made a stop at one of our outposts. My Soldier, PFC Gino Caprone and I saw PFC Orlando Gonzalez standing tall on his first mission outside the wire with his platoon. He was excited to see us and proud of his role on this operation. I loved that kid; he was always so happy and kind of goofy too. He always made us laugh. Gonzalez would run like a cheetah - he was lean and super-fast.

The three of us hung out at the FOB, but he and Gino were close in age and great friends. When I saw the convoy was getting ready to head out, we headed back to the vehicles. I turned to see the big smile on Gonzalez's face and wished I had snapped a photo of him, out there at his post, so proud to be on his mission.

We headed over with the convoy to an abandoned schoolhouse not far away to set up our outpost. It was a one-story building shaped like a *U* around an open courtyard where we staged the vehicles. We heard and felt a loud explosion, not far away. It was a big one. You could tell the enemy was all over the place. This was going to be a crazy op.

At night we found out four Soldiers were killed right by the area we had stopped by just hours before. One of them was PFC Gonzalez. My heart sank at the news. Not Orlando! We just saw him hours before. I remembered his smile, his happiness to be out there at that position. This sucked royally. There was an early morning mission, so grieving would have to wait. I had to tell Gino his buddy was killed. This was not going to be easy.

I walked over in the dark to the room where Gino was sleeping on the concrete floor on one of the Iraqi blankets. I put my hand on his shoulder and called his name. His eyes opened, but he was still sleepy. There were two other Soldiers in the room crashed out. I said, "Orlando's gone. He and three others were killed in that blast we heard earlier."

Gino's eyes got clearer, then he put his hands over his face. I continued, "We're moving out [on a mission] at 0600." I paused, then said, "I'm sorry, Gino." I left the room to let him grieve. That outright sucked. Bad news does not get better with time, but we still had missions to go on. I hated having to deliver that kind of news. That was the hardest thing I had to do my entire time in Iraq. I finished some prep for the mission, found some floor space, combat snuggled my M4 and closed my eyes for a few hours.

All that matters

Time had come for our civil affairs unit to rotate back to the States. We had been in country a year. I carried a sense of guilt for returning home in the middle of major operations while my buddies were still there fighting, not unlike when we left in 2004 while Fallujah was hot. Of course we had no control over unit rotations, but you can't help but feel like you should be there with the warriors you were just fighting with.

I was especially bummed to be leaving the 5-73 CAV. The last few months' experiences - save the hardship of losing fellow Soldiers - was awesome. I could not have asked for a more amazing soldiering experience. How crazy does that sound? Like, who describes it that way? But – it's true. I loved being a Soldier, serving on a team and supporting missions and operations. This is what I envisioned joining the Army and going to war was going to be like.

Our civil affairs' company had a formation for end of tour awards. All the team leaders and team Sergeants were awarded a Bronze Star except for one: me. It was one last way for my previous sucky leadership to stick it to me. The funny thing is it really didn't matter. Besides the fact that without a *V* for Valor, the award doesn't really mean much. They knew my level of dedication, and that I volunteered above and beyond for more missions, even the more dangerous ones that many backed out of – it even said that in my final evaluation. But more importantly, I know what I did on this tour and I slept with a clear conscience. Besides, I got something far better than an award. I earned the trust and respect of active duty infantry troops. The 4th ID combat patch and the 82nd Airborne Division combat patches the PFCs gave me from their uniforms still means more than any military award. The patches and an email I received from another troop a couple months later made everything worthwhile.

In the email, the Soldier thanked me for being strong in the middle of the operations we were in together. This was the Soldier's first deployment and he came in when the battles were hot and intense. He said my strength and calm during those days we were out in the fray together helped him through it. I was surprised and humbled to read his words. I hadn't done anything unusual, but I suppose you never know when your inner strength emanates and touches someone else who needs it. That he took the time to share that was very special to me.

DECADE OF DARKNESS

Yet the poor fellows think they are safe! They think that the war is over!
Only the dead have seen the end of war.

−Attributed to Plato

War was therapy for 9/11. But nothing prepared me for the real battle, when I returned home in 2007. Our unit flew back to Fort Bragg for a week of outprocessing. We turned in gear and went through various briefs. Of course, I did not know at the time, but a week of admin and then being released to the world you once lived in was like being ripped from a protective womb prematurely. Good luck "transitioning." Especially those that didn't have much to come home to.

I sat in the auditorium looking around. None of the hundreds of Soldiers sitting there had visible wounds (e.g., missing limbs, loss of eyesight, or burns). All those casualties had been evacuated from the theater. We were lucky, I thought. We survived without a scratch. But what I did not realize at the time was that we all brought the war home with us. Each and every one of us. Many would suffer from the invisible wounds so common to our war – multiple traumatic brain injuries (TBIs) – from blasts and bombs. Some would suffer from the horrors of war that manifested themselves in nightmares. Some would suffer from exposures to toxins that would make epigenetic changes to their DNA and affect hormones. I wonder now, 12 years later, how many that were in that room are still alive?

Two days after we got back, we were watching the news inside a trailer and saw a headline about a massive double-VBIED in As Sadah, at the compound we were just at. **Nine** Soldiers were killed when the schoolhouse collapsed from the blast. Oh my God... My heart sank and I felt sick. I literally felt sick knowing 5-73 CAV was right there. As soon as I got access to a computer, I emailed them but knew that communication lines on their end would be blocked until notification of kin of those killed in the brutal attack. I sat and stared at nothing. My body was here in the US, but my mind was still in Iraq. Empty.

As soon as we were released into the civilian wild, I got in my car and sped up I-95 north to visit Arlington National Cemetery to see PFC Orlando Gonzalez's gravesite before the cemetery was closed for the day. I needed to see him. I needed to connect with someone I was familiar with and trusted, and he was the closest one, even if his body was resting in peace six feet underground. I wanted to talk to him and say, "I'm sorry I didn't snap a pic of you on your first mission when we saw you at the outpost." I wanted to talk to him about the men we both knew, the brothers he served with that were just killed. I wanted to leave him my beige ball cap he loved borrowing whenever he stopped by my trailer. I needed to see him.

I was pulled over by a state trooper in Virginia. As he came over to the car, I debated telling him why I was speeding but couldn't formulate the words. Inside I was screaming, "I have to see him before they close!" but I felt muted. No sound came out. As he ran my plate, I waited in the vehicle and watched the clock, which seemed to move quickly. Time was passing and making the window narrow to get to Arlington in time. The trooper came over and issued me a ticket for reckless driving for speeding excessively. Defeated, I continued north on I-95 at the speed limit. The cemetery was closed by the time I rolled through. That ticket probably saved me from an accident.

My home unit in Staten Island, NY was always very good to me, and the leadership there looked out for us. I was offered the opportunity to attend desirable Army schools and put in my packet for E-7, but I didn't care about any of that. I wanted to deploy again and be in the fray, but I was not well and knew enough to recognize that going back over being less than 100% would be selfish on my part.

I was tired, and felt like shit, physically and mentally. I was disconnected from everything and everyone. I missed my friends, and they were not here with me.

Thankfully a retired officer at my reserve unit had reached out to me toward the end of my deployment. He was working as a contractor for the Navy in Virginia Beach, developing their civil affairs capability. Thanks to him I was able to get a solid job after returning home and relocated to the area to work as a military analyst, as a defense contractor.

It was the best bet I had going, allowing me to start a new life post-war, and in the midst of a divorce, as I adjusted to life as a civilian again, away from the fraternity and brotherhood I had been a part of for so long. I was an Army vet in a Navy town, in a new city where I didn't know anyone other than my coworkers, few who chewed the same sand I did. I felt naked and vulnerable without my M4 at my side. But working on a military base made me feel safe for part of the day, at least.

And this is when I found myself alone in my new apartment, alone with my pain and anguish, and alone with my thoughts. Despite the fact that I was resilient enough to not turn to the bottle of alcohol - the only thing I believed would temporarily remove the pain - I still had the daunting task of figuring out how to get through the pain.

Suffering in silence

War inside my head-can you feel it

–Suicidal Tendencies, "War Inside my Head"[7]

Everything was a struggle. I did not feel well. I was tired, my nerves were shot and I initially jumped when I heard loud noises - even the auto-flushing toilets in highway rest stops got to me. That was pretty funny actually. Thankfully many of the physical reactions dissipated over time. Two recurring nightmares played out in my head. I felt out of place everywhere and did not have anyone or anywhere to turn to. Who would I turn to? Who would understand? Half of my buddies were still in Iraq, the rest were scattered anywhere else in the country, but they were not in Virginia Beach. I missed my friends.

One night in desperation to reach out to someone, I called my soon-to-be-ex, the Marine. While I cried incoherently, he listened. We were currently working on a legal agreement to split our assets equally and civilly, rather than defer to a judge's decision. But he still took my call late at night. He didn't say much or have answers, but I could tell he empathized with whatever it was I was going through. I made it through another night.

Weird things were happening to my body too. I noticed a strange phenomenon. When I drank an ice-cold beer, even on a hot day, my fingertips would instantly turn icy to the touch. I couldn't explain it and it seemed harmless yet weird so I labeled it as a cool bar trick I could do. I didn't understand why I felt the way I felt, depressed, anxious and lonely – unable to find joy in anything. I learned the art of suppression of what I was experiencing in order to survive.

My parents had raised me to have personal accountability, to not complain about things or be a burden, and to solve problems on my

---

7 Suicidal Tendencies, "War Inside my Head," track # 6 on *Join the Army*, Caroline, 1987, LP.

own. In 2007 there were hardly any programs for returning vets. If there were, they were little known. So reaching out for help left you with one resource: the VA (Veterans Administration). I dared not reach out because I had a security clearance to protect to continue working as a defense contractor, and back then, in the early war years, it was believed your clearance would be jeopardized if you dared see or talk to anyone.

For the next ten years, I suffered in silence, living alone like a recluse in my small home in southern Virginia, save a couple break-throughs that I will delve into. I was highly functional – employed as a defense contractor for years, and eventually completed the master's degree I had started years prior, thanks to the Post-9/11 GI Bill. I never deviated from physical fitness, that had been a staple in life since high school, yet it wasn't enough to crush the darkness in my life for more than a few hours. I hid it well, lest anyone saw how empty my soul was.

I remember watching "experts" claiming that there were no weapons of mass destruction in Iraq and thinking, What the hell do they know? They weren't there. There most certainly were. I watched the narratives that were being pushed out by the media and in politics and felt even more disconnected from the masses. But the critical thinker in me was starting to realize the dangerous power of the promulgation of false narratives combined with the collective incli-nation of Americans to accept things at face value as we morphed into the sound bite generation. This troubled me.

I had mastered the switch of not connecting and not being emo-tionally tied to anyone. It kept me alive, kept me safe. It's what I knew. While watching the Angelina Jolie movie *Salt*, I related with the scene where her hubby was killed in front of her and she just drove on with the mission. I realized that was me. Unattached. Unfeeling. And that was dangerous. I didn't *want* to be that way, but I was.

Thankfully, I recognized that in myself and knew that if there was ever a way out, I should take it.

When you have been betrayed by your family - parents and spouse - **and** by your American military leadership, who can you trust? Your self-reliance is key to survival. You further develop a self-reliance like no other, letting very few people get close to you. I dated but connections were limited, kinetic engagements. I wasn't going to let my walls down, nor were they going to either, so a bridge was never built. No heart and mind campaigns developed, if you will.

Living in a beautiful beach area allowed me to connect with nature and the water, diffusing and relaxing. On the other hand, it further isolated me from humans. I felt like a fish out of water in social situations, so I usually avoided them.

It was hard to go back up North to see my family, so I hardly ever visited home. Ground Zero was still a giant hole. It was hard to watch and relive bad memories, remembering what I had seen on 9/11. The hasty metal fencing on the perimeter was replaced with a more attractive cordon, but the empty space was there. I still "saw" the mountain of debris that looked exactly what you would imagine in a war zone.

I spent most Thanksgivings and Christmases alone, and it sucked. I started to stress about holidays and long weekends because it outright sucks spending them alone when you should be making memories with loved ones, sharing normal human experiences with people.

I looked at the great and wonderful things Dad provided in my life and realized he was human and just doing what he knew to do because in his mind, he cared, and it allowed me to grant forgiveness and make peace with my father shortly after I returned from war. This was a huge blessing because unbeknownst to our family, a long decline of Alzheimer's started to kick in (none of us knew how long until we looked back and identified the signs). But even as it kicked in, I was able to create happy, loving memories with him that replaced the painful ones of the past. It allowed us to connect with few words and find peace between us.

Dad and Mom visited every year in the summertime for a few days, and it was always a great time, but oh so brief. Dad would inventory what I could not fix myself around the house, and we would start off with a trip to Home Depot or Lowe's. We would go sight-seeing and eat delicious fresh seafood, but then they were gone.

On one visit Dad said something that was a bit uncharacteristic for the advice he usually shared. He simply said, "Be happy. I want you to be happy."

I wish I had a stronger, closer relationship with him in the years after I returned from war. We made peace and were in the best terms, but I wish I would have had more memories with him, like fishing that we did when I was a kid. But outside the short visits, I was trapped in my world and did not get out much. I had created my own prison. I share this to encourage people to break down the prison walls you created so you can be free and have healthier relationships.

Outside of those few days a year, no one came by. A neighbor stopped by one time to borrow something and could not believe how immaculate my home was. It's not that I was a clean freak. I just did not have a whole lot going on in life besides work and working out, so my home seemed "untouched." **I wasn't living; I merely existed.**

I've no doubt that you think I'm off my head
You don't say but it's in your eyes instead

–Iron Maiden, "Still Life"[8]

8   Iron Maiden, "Still Life," track #6 on *Piece of Mind*, EMI Capitol (North America), 1983, LP.

Lethal formula and winning strategy

When I started writing this book, its focus was originally on veterans, but then it grew to something broader for several reasons - the first being that the problems vets face are not unparalleled in the civilian population. For the past few years, it has been accepted that the military veteran suicide rate is "22 a day." The reality is this is much higher given the fact that certain states do not report statistics and deaths are often miscategorized. For example, the "death by cop" scenario, where a veteran is violent and aggressive so the police are called to the home, resulting in a standoff in which the vet is killed for having taken the first shot, would not be considered a suicide – this happens way too often.

There is the "death by vehicle" in which the veteran drinks and drives (and is often on medications as well), crashes and dies. The point is that in these cases the vet was attempting suicide, but it was not classified as such. Additionally, if the family of the veteran does not want suicide on the death certificate, "natural causes" are often listed. All of these instances blur the true number of the veteran suicide rate.

The rates of cancer and suicide among veterans returning from war today and among the special operations forces are extremely high. Why and how is this possible?

Chew on this: psychiatry has never cured anything. Imagine a plan to flood the population, including children, with psychotropic drugs that have black box warnings for an increased risk of suicidal or homicidal thinking, feeling and behavior - knowing that statistically a percentage of the population would be affected by these "side effects." Think you would see an increase in mass shootings? What is going to be blamed here? The guns or the drugs?

Peel it back another layer: veteran suicide. Why is it that the veteran suicide rate is so high, yet the majority of veterans who commit suicide have not been deployed or in combat? The Department of Defense spent billions on prescription medications. Follow the money. Critical thinking is paramount in an age where information is weaponized. Ask questions and use the gray matter in your head.

I believe the combination of being doped up and disconnected are a lethal combination with a terminal ending, and both veterans and the younger generation of civilians are experiencing this.

Doped up is not about illegal drugs, though it can be. The "legal" opioid and psychotropic drug prescription problem in the veteran community and civilian population at large are at an all-time high. The disconnected reality stems from the world of technology and social media we live in, where communication and real relationships - areas that take work - are discarded for isolation and electronic/digital chirps.

When you come out of the womb with an iPhone in your hand, you enter an unnatural world. When you are on multiple medications, like selective serotonin reuptake inhibitors (SSRIs) with suicidal ideation side effects, you are living in an unnatural world. For many veterans the loss of purpose and identity further compounds the isolation. For many young Americans who grew up with technology, where their communities are based primarily online versus face-to-face, their social isolation is compounded. See, the plight between returning vets and younger generations of non-veteran Americans is not all that dissimilar, even if their life experiences are polar opposite. Older generations who are on the medications have had the opportunity to develop communications and relationship building skills before the internet age, so they have that leg up going for them. They likely experienced life that hardened them a bit too, before the emergence of safe spaces and pollical correctness.

I'll delve further into these areas, but for now I propose the path to being a functional and fulfilled human (again):

- Make the decision to change.
- Get off the drugs, prescribed or otherwise. (Consult with a physician first.)
- Invest in your health through nutrition.
- Get out of your comfort zone and grow.

Anyone can achieve this today, but only if they *choose* to do this. There are no excuses; it's ultimately your choice. Our country's GDP would be higher if vets and millennials embraced this. Our country would be winning more, and therefore, humanity would be winning more as well. As I've shown in my war examples, it's a lot better to be part of a winning team than a losing one.

Breaking the chains

I was tired of working for or with people who were toxic, no matter how good the salary and benefits were. I felt underutilized and, along with my coworkers, tired of the bureaucratic fallout that is inherent in government contracting. A great paycheck and a benefits package does not equate self-actualization or fulfillment. I wanted the freedom to work and partner with the right kind of people and not be beholden to a toxic environment. So in 2012, I stepped into the small business arena, in a climate that was not very small business friendly. I self-funded and started my first business - a service-based business solutions provider, leaving corporate America behind and never looking back.

Giving back

I stayed house-poor and lean, socking away money and continuing to invest and grow financially. It wasn't part of a grand plan. It was just what I knew to do and it came easy because there wasn't anything or anyone worthy of my spending money and time with during my decade of darkness. I followed through with the graduate program I had started between deployments, eventually graduating. So it wasn't the MBA I once wanted - a masters of business administration - but it didn't stop me from achieving an MBA - a massive bank account.

My business grew via word-of-mouth, and I scaled with strategic partners. I enjoyed working with small businesses helping them develop long-term strategies and using a root cause analysis and systems-of-systems approach to help them identify and solve problems. I nerded out on this; it does not feel like to work to be doing what you innately enjoyed doing. It was also extremely rewarding to create something from nothing and to provide value to the marketplace.

I started volunteering with the Patriot Guard, riding my motorcycle as part of the escort for fallen warriors. We would stand in a line holding American flags for the families or as the coffins came through. Some of the hardest ones were the young active-duty service members who did not live a full life. It was always a challenge but a very sacred honor to stand tall and strong with my fellow volunteers, not showing emotion and being strong for the families. I was deeply honored to be able to be one of the pallbearers for fallen veterans who were buried at sea.

Serving others unselfishly helped me feel better about my empty life. I spent my time, talent and treasure serving veteran nonprofits and helping fundraise tens of thousands of dollars for them. Having a mission greater than myself felt good. Being part of something that helped veterans, no less - many worse off than myself - felt great. It gave me purpose.

There were no shortage of veterans finding themselves alone with no fire team besides them, the proverbial fish out of water struggling with life, having unexplainable health issues. I could understand them and I wanted to help them. The downside to this is that I saw the dark and worst side of surviving veterans. That reality was not healthy for someone who was struggling, and there were some incidents that sent me spiraling further into a deeper depression, though, no one knew because I hid it very, very well. I had to. I hadn't found anyone since coming back from Iraq that I could trust.

Looking back, I am fairly certain that the charity and volunteer service I poured into and immersed myself in eventually brought about the karma that helped change my life.

MENTAL DISCIPLINE

Working on my masters while working full-time and part-time required discipline. My coursework was 100% online, so I did not have the advantage of a physical classroom with people to connect with. I sacrificed hours of time that could have been spent doing anything, from relaxing to enjoying life to living new experiences. Do I regret it? No. I felt accomplished for finishing a goal I set for myself. But more importantly, the process was invaluable. It forced me to focus and reprioritize things in my life.

The sheer act of following through, even when my heart was not into it, generated a deeper level of discipline. This was a very long test of endurance, given I started graduate studies in 2005 and completed in 2011. I did not choose to obtain a graduate degree for a career benefit, but rather for love of the subject matter (history, asymmetric warfare) and a purpose. What I realized later on was the process and journey that required a high level of discipline increased my threshold of non-physical self-discipline greatly.

When I started my first business and worked from home, people would say, "I don't know how you do it; days and nights sequestered while working on it. I could never do that." I realized that now that I was working on something I wanted to build, rather than on something I was instructed to read/research/write about; it was far easier. The process of working on my master's definitely helped me focus on being an entrepreneur. Not to mention the internal confidence of *knowing* I had the ability to complete a mentally demanding, dry, long-term journey as such. Many first-time business owners fold as soon as things get tough, or they do not have the long-term endurance to continue hustling hard to become profitable. I know I will always be able to see things through, even when they get tough. Even if I never used what I learned in my actual classes, I learned discipline and endurance by finishing my graduate degree.

SYSTEMS AND END STATE

Everything is run by systems. The quicker you figure out the systems, the sooner you can use them. Systems aren't always fair, but don't expend time fighting a system as an individual. It takes someone with a lot of resources to individually disrupt systems. A rare example is President Trump disrupting the traditional political system.

At a very individual level, having systems in your life helps organize your thinking and your actions. When you encounter a problem, it is a lot easier to troubleshoot a system and identify which process has failed than to blindly go into problem-solving mode. SYSTEM = Save Yourself Time, Energy and Money. The simplest way to start is to plug into an existing, proven system. People often try to create their own way of doing things, which is a great idea until actual execution. Leave complexity to analysis; the mark of a good system is that it is simple and duplicatable.

Our thinking shapes our actions. When you learn to ride a motorcycle, you are taught to physically look where you want to go. Your mind automatically steers the bike toward that direction. Apply that concept in life.

It is important to keep digging deeper to identify the appropriate goal to focus on. To identify the goal, you have to identify the desired end state and develop a strategy to reach it from there. You always need an exit strategy as well. I had a big goal once and achieved it only to realize that it was short-sighted and only "got me in the door." In retrospect, successful completion of the mission *once I got in the door* should have been the goal. As a result, I felt successful from the preliminary portion, but I did not shape my mind for success once the door was open. Things happened fast, and I failed. That is why it is important to have the *end state* in mind first. The end state will automatically encapsulate the preliminary goals.

Listen to the universe. I believe God helps those who help themselves, provided you genuinely take action to advance yourself, your goals and your dreams - and don't just sit there and think about it. Doors have opened for me that I was not looking for, but I knew enough to recognize these doors were linked in a similar pattern so that it was not a coincidence and worthy to acknowledge and pursue. That is the universe's way of pulling you along in a direction.

For example, let's say you want to focus on a singing career despite your natural talent for dancing. While you are struggling going to auditions, you keep getting paid gigs to dance and showcase your skill. You take the gigs to pay the bills but focus on your future singing career. Then one day you receive an opportunity to be showcased on a national platform. Some of your peers were selected for their singing skills, but you were selected for your dancing skills. You are concerned that appearing on this will brand you as a dancer rather than as a singer. On the other hand, the chance to be on a national stage can take you from obscurity to becoming a known entity, that can mold and change your life altogether, despite it not being in the manner you dreamed of. What do you do? I believe you roll with it. Keep moving! The doors will keep popping up in front of you. If you stop, so do the opportunities.

You can transition, scale and expand into singing later in life, but you may not get that national stage opportunity again. Accept that sometimes life's goals aren't attained in *your* timeframe but never quit on them, so you are more on track to get there.

Now, in the event you are inundated with the opportunities you want - for singing in this example – do you say yes to everyone? Do you risk becoming reactive or are you driving the train? This is why it is important to develop a system for yourself to help you determine the clear choice when faced with a decision. Having a system allows you to be strategic rather than reactive and helps you avoid "mission creep." Let your system be your guide.

An example of some systems for the singer:

-Assess. Every three months write down why you want to be a singer and put the date on it. Do not look at previous answers you have written. Store it in an envelope called, "My Why."

-Commit. After you write down your why, write down a concrete goal with numbers and a timeline of goals for six months, one year, eighteen months, two years, etc. In the singer's case, goals can be anything, e.g., sing on ten stages in six months, increase number of live performances by 10% each quarter, or earn $40,000 after taxes in one year, etc.

-Execute. The **only** thing stopping you from achieving your goals is you - specifically your level of effort. If you have identified that you average winning 3 out of every 100 queries you make, do the math to figure out how many queries you need to make in a day or a week to achieve your desired goal.

-Every year do the quarterly assessment. Then review the previous year's answers. See whether your answers have changed, if ever so slightly. Spoiler alert: people often give up on their dreams and goals once the harsh reality sets in. No shame. Be in the position to recognize that for yourself first in a legitimate, keeping-it-real way and save yourself time, energy and money.

Developing this simple system requires that you are truthful with yourself. No one else needs to see this list/system. But you have to be brutally truthful with your why and your priority elements. Having this system will remove the stress you would otherwise incur in life when being faced with a tough choice between different options. The system will make decision-making clinical and make you more decisive. Decisive trumps reactive. Give yourself the ability to see what you are doing from a level up - from the upstairs window, not the crowded sidewalk.

Still alone

I continued to struggle in silence, alone with that dark cloud coming and going. At the suggestion of a friend, I went to the VA once to speak to someone about how hard it was to cope and see if there were veterans' groups I could connect with. The doctor just typed away on her computer, then after a couple minutes looked at me and asked me if I wanted to be part of an experimental LSD program the VA was running. Are you freakin' kidding me??? I sprung up and said, "Hell NO! Forget I was here," then I beelined it to the parking lot. It was like a horror movie. Way to make me isolate more. I could just picture people in white lab coats sitting in a no-window room coming up with ways to experiment on vulnerable veterans without any form of oversight. I could not believe that just happened! Back to my safe house.

I went to a church service, but in the midst of the beautiful choir singing, I finally broke down, after years. Unbelievable waterworks. I was drained and extremely uncomfortable being vulnerable in public. I had a business to run, and who would "unsee" what they saw in a broken-down person? I knew I needed Jesus, but I didn't want any human to know. Back to my safe house.

Heavy metal therapy

Slayer makes the best war songs, literally. I would know! Thank you, Slayer, and other heavy metal bands, for getting me through soul-crushing moments in life - during my teenage years, through intense workouts, on the battlefield and during times when my thoughts were dark, raw, and full of anguish. Your music was my therapy for many years. Your music and primal screams were intense enough to keep me away from any chemical substance, no matter the pain I was enduring! The intensity of your music helped temper the intensity of life I was experiencing.

In 2016 I went to see Slayer at the NorVa in Norfolk, VA, for the first time in many years. Incidentally, it was also the anniversary of a significant loss for the special operations community. I ran into some veteran friends and paid respects with a Fireball shot to absent friends, a toast to our fallen.

The show was epic, exactly what you would expect from Slayer. Maybe it was because I hadn't seen them perform live in forever. Maybe it was because I had a lot of unresolved pain still inside me. Maybe because I had no cares to give. Or maybe it was the Fireball. Whatever the reason, I jumped in the pit. I had not been in a mosh pit in years. I had been built-up and broken-down since. I had seen evil, buried too many friends and life had kicked me in the gut a time or two. But I was 42 and alive and needed to celebrate in a manner I knew best. As I moshed and fell, got up, was pushed, was helped up, and continued around and around, over and over again to the epic live sounds of "Fight till Death" and "Dead Skin Mask," I released pain inside of me. If only I jumped in a pit in 2007. I didn't want this to end, but I was smart enough to have had my fill after six songs. I saw my friend, a Marine who had jumped in the pit as well. He grabbed my arm and pulled me out. Ahh, the power of Slayerrrrrr!

THE LIGHT: HEALTH FREEDOM

Food as medicine

In the '90s, I was skinny. Not skin and bones skinny, but thin. Barely over 100 pounds and most of it was big Jersey hair. In reality my body was in a catabolic state from lots of cardio from running and eating less than 1,000 calories a day. For years I ate Special K cereal with skim milk. I thought what I was doing was healthy. I *wanted* to be healthy, and during this time, that was one of the healthy narratives. This was before the internet age; the vast majority of information and influence was chiefly through television and widely-accepted branding. My stomach was a mess. I thought having bubble-guts was normal. Then I read an article about how many veterans returning from war were prone to metabolic syndrome - high blood pressure, high blood sugar and obesity. I did not know if I was prone toward this, so I continued prioritizing exercise.

In 2008 I started working out in a local functional fitness gym. It looked like CrossFit in terms of equipment used but was more mixed martial arts-based, versus gymnastics-based. Virginia Beach had a very close-knit CrossFit and functional fitness community; we often supported each other and worked out together frequently, especially during Hero workouts which were dedicated to fallen warriors, many of them from our community, given the proximity of Navy SEALs' East Coast–based teams. January 15, 2011, I attended a seminar at CrossFit Takeover. It was called, "The Paleo Challenge." I knew little about it but knew that many of the top athletes and trainers in our community who had high-performance lifestyle habits were involved in this "paleo," and I wanted to learn what this was all about.

To say it was an eye-opener for me is an understatement. I learned the principles of eating what was on the Earth during the Paleolithic "caveman" Era, before the advent of agriculture, which only started about ten thousand years ago and has given way to a lot of inflammation and disease in the body. A lot of this was revealed in a book by the man considered to be "the father of the modern-day paleo movement," Robb Wolf, in *The Paleo Solution: The Original Human Diet*. I learned how he was able to reverse his health conditions through reversing what he was eating. Food as medicine to heal the body naturally, what a concept. I learned about the US government's food pyramid lie, the perils of milk, grains, refined foods and genetically modified organisms we were eating - "Frankenfood." I was deeply intrigued and enrolled in the paleo 30-day challenge.

That afternoon I went home and threw away the boxes of Special K cereal, and a half gallon of skim milk. Not stopping there, I ruthlessly went through my pantry and threw out everything that was highly processed, full of preservatives – which was mostly everything. I had grown up with the best Italian food all around me in New Jersey. Pasta was a food group! I dumped the pasta. I remember being alone in my kitchen that Saturday night, thinking of the lies about nutrition the government - and subsequently, school - had fed (pun intended) us over the years. I was angry. Thirty-eight years of my life I had been eating mostly crap. Grr!! Maybe this is why I felt like bad no matter how much I worked out. Maybe it was the gluten, maybe it was the glysophate.[9]

---

9   Samsel, Anthony, and Stephanie Seneff. "Glyphosate, pathways to modern diseases II: Celiac sprue and gluten intolerance." US National Library of Medicine National Institutes of Health, December 2013. https://www.ncbi.nlm. nih.gov/pmc/articles/PMC3945755/#CIT0229.

But I was also faced with a challenge: Where am I going to eat now? Back then no restaurants offered paleo options; "gluten-free", "non-GMO", and "organic" were not buzz words. My idea of cooking was using a microwave or tuna fish from a can. Nothing was going to stop me from cleaning up my diet. I had one choice to get there: start cooking my own paleo meals. I went to the grocery store and started focusing on foods on the perimeter of the store - fresh meats, fish, fruits and vegetables - skipping the aisles of highly-processed, packaged foods. I discovered things I never heard of - starfruit, dragon fruit and spaghetti squash. I started experimenting with herbs to make my own seasonings that did not contain harmful additives.

The results were immediate and uncanny. My energy levels went through the roof. After three days of eating grass-fed beef, bacon, vegetables and almond butter, I felt like a rock star!! As time went on, my gut wasn't a mess. I remember eating a strawberry after three weeks of clean eating: since my taste palate had changed after removing junk from my diet, it was the sweetest strawberry I could remember!

The paleo challenge had a built-in "cheat meal" allowance, which I never partook in. I was way too disciplined to cheat myself. This was a lifestyle choice, so it was a no-brainer to commit, make no excuses and go all-in. I literally obsessed. I found the International Paleo Movement Group (IPMG) on Facebook, ran by Karen Pendergrass of the Paleo Foundation and learned more information.

I started posting my meal creations on social media and was encouraged to write a cookbook. I did not have any interest in doing so, but I did see the value in sharing what I was cooking and doing so it could help others who may be looking to eat cleanly. Then I realized that if more people embraced paleo, restaurants might start offering healthier options - foods that were not cooked with hydrogenized oils would be a great start. I created a Paleo Calendar each year from 2012–2014, with 100% of the proceeds going to a veterans' charity. It was my hope that people would keep the calendar on their refrigerator and start making better lifestyle choices. If you look up Paleo Calendar on Facebook, you can still see photos of the amazing meals I made up before the market became saturated with paleo cookbooks.

Brain on fire

A funny thing started happening. I did not understand it at the time, but my brain started improving. I experienced better focus and less brain fog - probably the first time I felt like my brain was "on fire," working at a very high level for a sustained period of time. I was finishing up my last semester for graduate studies while working full-time and coaching functional fitness part-time. Two separate professors commented that my papers were much better than anything I had written previously. They did not know about my health journey, so I noted this feedback as a substantial indicator that this "paleo thing" was working really well for me. Fascinating.

What I understand now and can explain is that this was probably a two-fold benefit of the paleo diet: (1) my brain was operating more clearly having removed inflammatory foods from my diet and (2) my brain was benefitting greatly from the addition of medium-chain triglycerides added to my diet, from coconut oil. I was fast-tracking toward a bulletproof mind, literally. Since I obsessed with the paleo lifestyle, my body became keto adapted, and I often entered a state of ketosis, with numerous benefits that are commonly understood today.

Another phenomenon I noticed was the social aspect of being so easily disciplined about clean eating. Sure, I was that "guy" that took forever to order at the restaurant, and still am. Some of my friends did not like my discipline. At a birthday party, one friend got frustrated that I would not eat the piece of the chocolate cake in front of us and asked, "How could you not eat even a bite??" as she ate her piece in frustration. I looked at the cake and saw dead food - white, "enriched" flour laden with high-fructose corn syrup. Okay, yes, on the surface it looked yummy - and for three decades, I would oblige - but knowledge is power. Now that I knew what was in it, why would I touch it? I had no desire for it even, as my body had adjusted to eating delicious, nutritious food, not filler.

I learned about the power of refined sugar and how addictive sugar was. Sugar comes in all forms and is in everything. I started looking at people who were addicted to sugar and saw a parallel with their moods and overall performance. More sugar = more emotional hot mess. Cray-cray and sugar addict go hand in hand. I am not saying the world needs to go paleo, and I acknowledge everyone's body and DNA are different. But don't complacently accept the status quo and eat everything you are given, so to speak. I am grateful to have found what works best for me and haven't looked back.

Proof of concept

I saw an opportunity to test the impact of paleo on my physical performance. I also wanted to test the impact of the overall strength and conditioning from functional fitness in an endurance run. So I signed up for the Yuengling Shamrock Marathon two months after embracing the way of the caveman.

In the two previous marathons I ran years prior, I "carbed up" the traditional way with pasta the night before. So now I needed to figure out what to eat the night before, and what to snack on during the arduous run. I knew I wasn't going to break any speed records here, but wanted to be prepared with a paleo snack for a boost. I gathered the starchier of the paleo foods and concocted some kind of mush consisting of two boiled sweet potatoes, coconut manna and three plantains to eat the night before and in the morning. Energy!

To replace the GU gels I would have previously carried on me for fuel on the run, I found a squeeze-pack of pure almond butter. As necessity is the mother of invention, I created and baked "paleo balls" - which included almond butter and chia seeds, which helps store water in the body.

I wasn't afraid of personal failure; if it were a team, I would have felt differently. I can't let a teammate down. So what was the worst that could happen? If my body shut down at mile 20, at least I would know I could make it to mile 20. Mentally, what was there to be scared of? A little pain and soreness? I wasn't going to die. I wasn't being shot at. I tuned everything out and pushed through with my favorite tunes, like Fear Factory's "H-K (Hunter Killer)" and the Misfits' "Die, Die, My Darling".

Though it took six hours, I completed the 26.2 miles without injury, fueled by paleo-keto food, wearing Vibram five-finger "toe shoes," having only done a nine-mile and two six-mile runs in the weeks prior to the race. Yes, I realize people can speedwalk a full marathon in that time! But most importantly I knew right then and there that if I could run that kind of distance fueled as such, I could eat and survive this way forever. Boom. My personal proof of concept of an unconventional trifecta of training, diet and footwear was successful. My physical and mental threshold were further deepened. Win.

Clean eating 101 makes sense. You can start by avoiding or removing manmade "foods" ("Frankenfood"), processed, packaged, and refined foods, and anything with fillers, chemicals, and dyes. Strive for real food, fresh food, locally-grown, and organic food. There are many inflammatory foods. Avoid GMO foods (genetically modified food), and hydrogenated oils. Remove questionable foods from your diet and see for yourself if you feel differently. Pro-tip: If you look at the label on fruits and vegetables, if it starts with a '3' or '4' it is conventional (average); if it starts with a '9' it is organic; and if it starts with an '8' it is genetically modified.

OUTGROWING THE TRIBE

When you are an early adopter of something, some people who do not understand will try to bring you back to their world because you are ruining the vibe of the tribe! For some people, your growth - and happiness - serves as a mirror to their stagnancy and unhappiness. The best thing to do is to limit association with these people or move on from them.

People in your life who may have been your friends or even mentors may not know what to do or how to respond to you when you have "outgrown" them. People who may be super confident in one area but have underlying insecurities suddenly experience inadequacy and even cop an attitude to you. How dare you grow! Key lesson here: surround yourself with secure growth minded people. Not only will that feed your growth, but these people will likely celebrate your wins and support you.

GUT INSTINCT

Sometimes the best deals are the ones you don't say yes to. I've had opportunities to scale my B-to-B business over the years but declined for various reasons, all ultimately stemming from a gut feeling about an aspect, be it market conditions, the leadership team, or the return on investment. I viewed it from a long-term lens. Looking back, I am glad I did not go for the quick win because my intuition eventually proved valid.

An example of this was a potential business partnership with a logical vertical. The owner was very aggressive about pursuing collaboration. The numbers looked attractive, but the trust was not there, and I did not proceed. Not long afterward, shenanigans with another partnership he formed came to the surface - and there was litigation. I wondered if he had thought I was an easy target to dupe. I thanked my gut instinct for protecting me from what would have been a bad business decision. So while I did not have as many short-lived wins as I could have, which may have advanced my business's growth, I avoided costly pitfalls that could have stunted it.

Speaking of gut instinct, did you know that scientists have linked the gut to the brain? Brain issues often stem from gut issues, so I will take it one step further and suggest you choose partnerships in life and in business with those who have a healthy gut. Think about it. And heal your gut for better decision-making. Heal the gut, heal the brain, heal the mind. Happy gut, happy life. Simple.

Because my brain-gut connection became healthy and optimal, and I was already growth minded, I met a thriving international network of millennials with committed partners who developed a system to empower and redefine their future through an online business platform. Thankfully I put my ego aside and learned exponentially from people much younger than myself who had figured out a pathway to wealth and success I had not yet explored.

With a small investment, I went from small business owner to adding an international business to my portfolio in a short period of time. The 10% annual compounding component made this a no-brainer. I was grateful my mind was free and clear to recognize and seize the right opportunity at the right time. This is still a project I work with steadily and would love to provide the same opportunity for anyone looking. For more information, go to https://americandreamthebook.com/timefreedom.

GUT CHECK

Are you really free? You cannot attain freedom by being a slave to your emotions or ego. You can have wealth, but if you are concerned about what people think of you, are you really free? Think about it. You cannot control your emotions if your health is a mess, because poor gut health will influence your brain health, your outlook in life, your mood, and your ability to think clearly and decisively.

Freedom definition:

- Learn to control your emotions and your mind
- Optimize your health
- Learn to build your net worth

So, improve your health, improve your outlook and improve your net worth!

Saving Michael

Michael was an Army veteran with multiple tours of combat. One night he texted me from his girlfriend's home halfway across the country. The text was quite ominous, so I called him right away. I could hear his girlfriend's young children in the background and listened as he babbled incoherently about his demons. I could tell he was very drunk. His girlfriend texted me that he had moved to the garage and was playing with his gun... OMG! I texted her to get the children and get away while I kept talking with Michael. At the same time, I messaged some mutual veteran friends who lived closer to Michael to see if they could get over to him and diffuse the situation. Calling the police would only have escalated the issue to a potential suicide-by-cop scenario.

My heart was beating as Michael talked about some very dark things. I was concerned that our mutual friends showing up may alarm him, so I kept him talking so he wouldn't hang up. What do you do in a situation like that? When he said he heard the sound of motorcycles outside and was alarmed, I said, "That's Adam and Jeremy; they just wanted to drop something off for you and the kids...." I held my breath not knowing how this was going to go down, praying it went well. Michael stopped talking and I could hear him moving around the garage. Finally I heard voices talking. It was not intense. Thank God ..., I sighed in relief. Michael was happy to see them and they weren't going to leave him alone. They stayed with him all night.

The next day I asked Michael to consider getting treatment at a brain treatment center to treat TBI from multiple roadside bombs when he was deployed. Thankfully he obliged. They used nonpharmacological, noninvasive treatments to heal the injured portions of his brain. He also underwent hyperbaric oxygen treatments, which helped increase the blood flow in his brain. Today he is married to the girlfriend that stood by his side and lives a good life.

Michael recovered from the worst moments of his life, but so many others did not. Of Mice and Men's song, "Never Giving Up" captures the anguish in my heart during the many times over the years, when military brothers went into their darkest depths of hell in their post-war lives. Sometimes we - the veteran community - could help them, sometimes we could not.

Functional medicine

Through the work I was doing with veteran nonprofits and my outright curiosity to research and find answers, I started learning the science behind why I felt like crap when I returned from war and had headaches, depression and anxiety. This was key to my realization that I had the power to change things if I knew was their root cause.

I inadvertently learned that the Mobile Multi-Band Jammer (MMBJ) anti-bomb device behind my head in the turret was equivalent to sticking my head in a microwave. I wondered if that was why I had so many headaches when I returned in 2007, or whenever I used Bluetooth devices.

I met Geoff Dardia, an active duty Army Special Forces NCO. Geoff was like many other veterans, "banged up" physically, mentally and emotionally. Facing divorce, medical retirement, deep depression and suicidal ideation, he refused to accept standard of care and refused to believe a large portion of his symptoms were "all in his head." After countless hours of research, Geoff was able to identify the health effects of his operational environment, detect the damage caused by them (root causes), properly diagnose them, then treat them with functional medicine care outside the conventional military medical system. The program he founded, the SOF (special operations forces) Health Initiatives Program within the nonprofit, Task Force Dagger Foundation[10], has provided direct assistance to hundreds of special operations forces by providing personalized medicine in the form of functional medicine - nonpharmacological alternatives, such as hyperbaric oxygen treatments and noninvasive neuromodulation for the brain.

The program uses a network of functional medicine institutions, such as the Cleveland Clinic and advocates for functional medicine physicians like Dr. Gabrielle Lyon[11], who leverages evidence-based medicine with emerging cutting-edge science to restore metabolism, balance hormones and optimize body composition - with a practice that also specializes in the SOF community.

Every action has a reaction. I learned this all too well. We're humans, not machines, and need self-care to keep running smoothly. I'm not mechanically inclined, but even I know to take my vehicles for an oil change and multi-point inspection every 5,000 miles (thank you, Dad). Looking back now, it's no surprise that all the health issues I experienced after a couple tours in Iraq happened afterall. There are way too many other service members who have spent much more time runnin' and gunnin' than I did who haven't received the self-care message. This applies to everyone who finds themselves in a high-stress environment or job.

---

10   https://www.taskforcedagger.org

11   https://drgabriellelyon.com

Geoff Dardia is leading the way in creating a model for change in the system, making the case for baseline testing for service members before every deployment and for successive post-deployment testing. This will allow correction of deficient/damaged areas and rehabilitation, before the service member deploys again. This in turn increases readiness of our warfighters and will lower the suicide epidemic of veterans during and after service.

I was tested by the VA and found out I had heavy metals in my body, likely from the various toxins I was exposed to in Iraq from the large burn piles, where everything from batteries to medical waste were destroyed, or "burn pits" as they are now called. The VA tested and confirmed a neurological issue I was dealing with, but does not link it to war. I started to understand that weird "bar trick" and the numbness issues I noticed as far back as that cold winter in Fort Bragg. I also learned that what everyone is quick to label as post-traumatic stress is also the result of blast injuries and toxins that affect your hormones and cause epigenetic changes to your DNA.

Look up "Transgenerational Epigenetic Inheritance" in which damaged genes are passed on to future generations. Children of Holocaust survivors experienced this, much like many of the offspring of veterans returning from battle today - where autism rates among their children are high.

After realizing that there was nothing inherently wrong with *me*, rather the damage was in my blood and nerves - not my mind, though it affected my mind - I became extremely focused on trying to remedy this. I felt cheated of many years of life and wanted to find answers, implement them and help others do the same.

Ground Zero and burn pits

There is a common link between service members, returning from Iraq and Afghanistan, and first responders from 9/11. When the Twin Towers came crashing down, it released toxins all over the area in that familiar "white dust." I remember a report that the EPA had tested the air around Ground Zero and said it was fine. I also remember the head of the EPA resigning shortly afterward.

The mercury alone in all of those fluorescent lights in the offices was extremely toxic. Eighteen years later, we now know that the levels of toxicity were extremely high, and first responders and workers who sifted through the mountain of rubble – often without masks for days, breathing it for weeks and months - have been dying from cancers related to toxic exposures at the site. Some cancers take years to manifest themselves, which is why we have been seeing so many deaths now.

During this writing, the 9/11 Memorial Glade opened as a tribute to those who have died as a result of these toxins. Part of the inscription[12] reads,

> *This Memorial Glade is dedicated*
> *To those whose actions in our time of need*
> *Led to their injury, sickness, and death*
> *Responders and recovery workers*
> *Survivors and community members*
> *Suffering long after September 11, 2001*
> *From exposure to hazards and toxins*
> *That hung heavy in the air*

---

12   9/11 Memorial Glade, unveiled May 30, 2019, inscription carving, New York, NY; view inscription online: https://www.911memorial.org/memorial-glade.

So while the Veterans Administration has a Burn Pit Registry, it has not officially recognized that the toxins service members have been exposed to from burn pits caused these chronic diseases and cancers. Suicide and cancer are killing more veterans than the number of veterans who have died on the battlefield. Service members have died from exposures to the burn pits.

Much like Agent Orange in Vietnam and the Gulf War syndrome that affected thousands of veterans, the government tends to acknowledge the problem years afterward, after so many veterans have already succumbed to illnesses related to exposures and died.

In addition to burn pits, veterans have been affected by anti-malaria drugs, controversial Anthrax vaccines and many other environmental factors - even the leaching of the BPA-ridden plastic into the bottles of water we drank that were sitting in pallets in the hot desert sun.

BE YOUR OWN ADVOCATE

It is no wonder veterans coming home feel like crap! Their bodies have absorbed what was in their environment, and they've been deployed in some of the dirtiest environments. Anyone - veteran or not - that lives or works in a toxic environment is going to be subject to health issues from environmental factors. Lifestyle choices (occupation, poor diet, lack of exercise, poor sleep, smoking and drugs) and environmental factors (air, water, soil, chemicals and pollution) affect us all.

The decision to do something once you have information is up to you. Advocate for yourself and your well-being. This means, ask questions, demand answers, and if you do not receive answers, find new or alternative resources. Invest in you!

Cellular health

All of this knowledge was a big clue that led me to believe repairing the damage from the toxins in my body could reserve or slow down the symptoms. The thing is your DNA can only split so many times before it's unsalvageable and damaged, creating the conditions for cancer. I learned about foods that support telomerase, which activates replacement/elongation of the protective telomeres at the end of chromosomes. I found the antiaging and reverse-aging benefit fascinating as well. While I had embraced the clean eating lifestyle, I needed to look into natural supplementation to get the optimal nutrition my body needed to heal more. I wasn't getting enough of these nutrients from available healthy foods, especially considering how our soil in the US has been depleted of nutrients and sprayed with toxic pesticides. Wellness became my focus and obsession.

Superfoods, superhuman

I looked into cellular health and started learning about nutraceuticals - powerful foods that help the body heal naturally. I started superfood supplementation to increase antioxidants to battle the oxidative stress and free radicals in my cells - and defend, maintain and repair them.

One day a friend I knew and trusted introduced me to high-quality blend, made up of superfoods I never even heard of from around the world, full of polyphenol-rich antioxidants and the mighty antioxidant, astaxanthin. Many of the superfoods did not grow in the United States because we do not have the climate or terrain they need. I started researching some of them and was blown away at the health benefits. I didn't have anything to lose, and it met my picky clean-eating standards, so I gave it a shot.

I woke up on the fourth day of taking these supplements and almost cried; it was so surreal. For the first time in years since returning from Iraq, I slept so deeply and peacefully and had good dreams! I had completely forgotten what it was like to dream. It's like that part of my brain had been shut off. But I was finally able to reach deep, restorative REM sleep, to the point I could dream again. Sleep became beautiful! I felt so much more rested and productive now that I was getting enough of the right kind of sleep. This was huge.

I couldn't believe how quick and powerful the results were. As I continued taking the various superfoods, more good things started happening. My energy kicked in to a whole other level, despite the lack of caffeine and stimulants. I actually felt less of a need to drink coffee in the morning. But the most powerful change I experienced was the dark cloud that loved to visit me finally went away. My depression was dissipating! I couldn't stop smiling! I became that annoyingly happy person, and it was glorious!!!

No one but my friend who introduced me to this knew I was taking it, yet people started to say things like, "There is something different about you." Wow. I was so intrigued I started studying the various superfoods more in-depth and believe the one that passes the blood-brain barrier was what my brain needed to start healing.

Convinced that Mother Nature has been a powerful force vastly underestimated, I started using a bioavailable liquid superfood to boost energy and increase my body's natural nitric oxide production. What I did not know at the time was the lead in my body was inhibiting my body's ability to produce nitric oxide, the signaling molecule of life discovered by Nobel Prize–winning scientists. What I did notice was an unbelievable increase in energy and more notably, my headaches went away with a few drops!

This was a big win because I knew the dangers of nonsteroidal anti-inflammatory drugs like aspirin, Motrin and Advil. As I read further about it, I learned it was backed by science with an independent study showing it had the same effect on pain as morphine but without the side effects. This was absolutely mind-blowing!

I met integrative health doctors who were prescribing this for their patients and who had seen significant health results. Given society's ills and pains and reliance on habit-forming opioids, I believed people needed to learn about this ASAP. So I bought international distribution rights and made it a quest to share this information with those who could benefit greatly.

People who already engage in high-performance lifestyle habits understand the concept of "nutritional psychiatry" because they have experienced it firsthand and understand it is a no-brainer. Of course, it is met with resistance. Using critical thinking, consider what entities would not want to see people healing their bodies and minds through nutrition, overcoming common health issues once believed to be treated only by medications? Do the math and follow the money to find your answers. Who stands to benefit from keeping Americans medicated, with side effects?

In August 2019, "A judge in Oklahoma singled out Johnson & Johnson, ordering it to pay the state $572 million and ruling that the company should be held responsible for decades of opioid addiction and the thousands of overdose deaths in the state."[13] For Americans' sake perhaps these rulings will help stem the tide of the pill-popping prescription drug problem in our country.

I have no doubt that my now eight-plus years of highly-disciplined clean eating, proper REM sleep, intentional lifestyle improvement, and three-plus years of antioxidant-rich superfood supplementation has helped heal my body and mind - and even slowed down the aging process. It's allowed me to sustain health freedom - the absence of medication. I have been able to "manage" the neurological issue as well.

---

13   Thomas, Katie, and Tiffany Hsu. "Johnson & Johnson's Brand Falters Over Its Role in the Opioid Crisis." *The New York Times*, August 27, 2019. https://www.nytimes.com/2019/08/27/health/johnson-and-johnson-opioids-oklahoma.html.

Who knew superfoods were key to becoming superhuman?! On top of how amazing I feel on a daily basis, my lab work supports these results. It's great to be 45 years old and on no medications, but so discouraging that this is far from the norm in America. I've been successful in using nutrition to heal my body and mind and know that this knowledge can help others who want to heal too.

If you would like more information on these superfoods, go to https://americandreamthebook.com/healthfreedom.

You, too, have the power to reprogram your brain through the addition of nutrition and the elimination of inflammation. Health freedom for the win!

The Miracle

Mom was faced with tough choices when Dad was deteriorating. Anyone who has seen a loved one suffer through the advanced stages of Alzheimer's knows exactly what I mean. She could have put Dad in a nursing home and visited him, leaving the tactical care to trained professionals. She considered all options but decided to take care of Dad at home. It was very, very heart-wrenching to see the transformation resulting from this terrible disease.

I drove up and down I-95 to spend as much time as I could with Dad as he declined in health. It was the second time in my life I had that feeling of helplessness, of not being able to do anything to reverse what was happening. Right after Thanksgiving, he went to the hospital for health issues, as is common for people struggling with Alzheimer's. When he returned home, he was weakened greatly and did not eat for three weeks. He wouldn't drink water but existed only on half a glass of a dark cola each day - something I knew wasn't helping. He mostly slept but absolutely refused to eat anything. He needed nourishment.

One of the proprietary superfoods blends I had been taking came in liquid form. I offered it to him, but he refused to sip it from the pouch, so we decided to pour it into the cola so he would get nutrition. After three days of this, he began to eat again! I was almost in disbelief. While I knew the superfoods were powerful, I needed to know the science behind which one was responsible for this. I started digging deep in research and found that one of them was known for opening up suppressed appetites. I'll never know for sure which one(s) in the amazing formula helped most, but I do know that Dad went from three weeks of not eating to being with us for another seven months because his body was getting nutrition to heal a part of him. Though this was not a joyous Christmas, I believed I witnessed nothing short of a Christmas miracle.

As hard as it was to watch how hard Dad's deterioration was on Mom, I was inspired by her strength through it all. She moved on autopilot, tending to Dad. She knew this was a better choice than leaving him in the care of a business. She knew he would rather spend his last moments in his home that he loved. Even though the day to day was brutal on her, physically and emotionally demanding, she loyally and lovingly did all she could for him. I will never forget what I witnessed, and thought what a remarkable superwoman she was, in a world where convenience often trumps doing the right thing.

As I sat next to Dad, I reflected on how growing up he had provided an environment that enabled me to dream. He set the bar of expectations high and inspired me. He led by example. He parented and provided and cared deeply about our family's welfare. He did the best he could.

The hospice nurse said patients whose organs and systems are starting to shut down, and may be in a coma-like state, still have two senses that are still strong - the ability to hear and the ability to feel.

Though he looked like he was sleeping, I patted my Dad's arm and spoke to him. I shared memories that we had enjoyed together, reminiscing on good times. And then I said, "Daddy, remember when you told me to 'be happy'? I AM happy! I finally found happiness.

Life is so much better! I'm so thankful you said that to me and that I finally found it."

And then I saw a tear roll from his closed eyes down his cheek. He had heard me, even in his state. I teared up but kept my hand on his arm. He couldn't open his eyes or respond otherwise, but I saw his response. This was so beautiful, so powerful, so meaningful, so healing. The love I had for Dad and the love he had for me was not questionable. I felt so connected to him. I felt so lucky to be by his side, and to share this. I had no idea if this was a common occurrence at end of life, but I felt very grateful to God for this powerful blessing.

I was grateful to able to be there with Dad, right at his side when he took his last breaths. And I felt a calm at that moment - indescribable in words, but beautiful.

Even though we didn't have an actual conversation, between the turbulence of my life postwar and the onset of his Alzheimer's, a conversation was not necessary for us to resolve our issues and find peace. I experienced this even after he was gone.

I saw reminders of Dad in natural occurrences - things that uniquely reminded me of him and let me know he was watching over me from Heaven. One of them was quite surreal. A month after Dad passed, I was returning from a business trip and was on a plane headed from Dallas to Washington, DC. Dad was always fascinated with space. He was proud that the Apollo landed on the moon on his birthday - the same year he moved to the United States. We watched the moon and the stars through the telescope as a kid, and he would tell us about the wonders of the universe.

The flight I was on happened to be on the same day as the solar eclipse. The pilot announced that we would be nearing the changes in the sky. I gazed at the majesty of the universe through the airplane window, watching the solar eclipse from 35,000 feet up in the air. It was beautiful and amazing. I felt lucky to have been at the right place at the right time and thought of Dad, who would have enjoyed this amazing view. I felt closer to God, and felt a peace, knowing although Dad was gone, he was always around and he would let me know through ways only I would recognize because I knew him so well.

BEST SELF, BEST LIFE

The strongest oak tree of the forest is not the one that is protected from the storm and hidden from the sun. It's the one that stands in the open where it is compelled to struggle for its existence against the winds and rains and the scorching sun.

–Napoleon Hill

Mindset is everything

Don't waste your time always searching for those wasted years
Face up, make your stand

–Iron Maiden, "Wasted Years"[14]

I have two groups of war buddies: Warrior *A* and Warrior *B*. Both groups are comprised of tip-of-the-spear badass elite warriors who have done some of the most incredible things on the battlefield, or even in training, to get to that point. Both groups transition out of the military. The Warriors in the first group take their skills and abilities to transition to a job or business. They are already growth minded and able to thrive in any environment. While the Warriors in the second group have all the same awesome skills and abilities the first group does, they hit that brick wall and shut down upon transition. They are no longer on top of their game because they are not literally part of the tribe they were in for so long and are entering the next phase.

---

14   Iron Maiden, "Wasted Years," track #2 on *Somewhere in Time*, EMI Capitol (North America), 1986, LP.

The second group of warriors adopt a fixed mindset suddenly where they once were open to doing what they needed to do to grow and become the best of the best in uniform, but now they're fish out of water. Here's the thing: I'm not talking about injuries, exposures etcetera being a factor because all these are factors for both groups of Warriors. So take that variable away; what I'm talking about is this other group comes back and they have a fixed mindset toward "civilians." They are a shell or they become a shell of what they once were.

Same training, same injuries, but what sets them apart it is their mindset. They have a fixed mindset, and basically, since becoming top of their game in another tribe - in this case, in the military, and for some, in an elite unit - they believe they can't compete in any other tribe, like the civilian world. This is because they have a fixed mindset. If they had a growth mindset, they would learn what they needed to learn and it would be a quick learning curve because they were already prone to achieving much very quickly.

I struggled with transition; however, ultimately I did not let that stop me from figuring out how to eventually crush life. And I was not even an elite Warrior. Yes it took me time, and yes this is one of the reasons I'm sharing my story. No, I'm not writing this solely for the military; however, with the rising suicide rate, there is certainly a sense of urgency on my part to address these issues. This Warrior group is but one example that can be applied across the board. I believe this extends even past American citizens. Maybe you live in an oppressed country. Maybe you're in a country where you don't have access to YouTube and all the free videos of mentorship. These tenets are universally written, and you can apply them in your environment.

Financial therapy and time freedom

You know what else helps get you past things? Living well. Making money. Having money. Having money that grows while you sleep. While money doesn't buy happiness - that comes from within - money can buy resources you need to give you the time freedom to spend your precious time on Earth with loved ones or doing what makes you happy or fulfilled. Why are so many people allergic to money?! Money gives you the ability to yield influence or make change. If you are a good person, more money will make you a better person because you can do more good.

Money is up there with oxygen in terms of importance. This message is not taught in schools or in many homes, but it is reality. Check out Robert Kiyosaki's *Rich Dad, Poor Dad* to learn more about the cashflow quadrant.

I found a natural fit in business – where everything is transactional, numbers driven (sales, revenues). "How is your business going?" you ask. "Oh, it's feeling great *today*—" Not! More like, "It's great! 2nd Qtr. profits are up 8%!"

Expectations are spelled out in contracts. When a business relationship ends due to anything negative (e.g., lack of proper fit, lack of performance/deliverables, change in requirements/capabilities, etc.), it is simply "business" - it's not personal.

In business you have to thrive to survive, which means you have to adapt and constantly evolve in today's competitive market. He who adapts first is first to market and generally stays ahead of the game. Drive that *S* curve! Business owners must have tenacity and a penchant for growth, which makes them some of the most interesting and attractive people in the planet.

Though I did not know it at the time, despite control by Dad, it was his lessons about financial literacy, the growth environment he provided, the nurturing he provided for my curiosity, and the example he provided for me as a kid that enabled me to achieve a financial milestone, well ahead of the average American. This eventually led to time freedom to live an intentional life by design, and spend it with whom I choose and how I choose, not with whom I need; I'm not beholden to anyone or anything. Freedom much?

Even by 2019's standards, that's probably the best definition of empowerment. Empowerment, the American dream! The best news is, there is nothing stopping you from achieving this a lot earlier in life. What is holding you back?

You may have heard the average millionaire has seven streams of income. There is no shortage of residual income generating opportunities in the post-turn-of-the-century era marked by the information age and globalization.

For years I followed the traditional pyramid of having a job, with a boss, who had a boss, who in turn had a boss, exchanging my time for money and selected benefits. It didn't matter if I was the one who showed up the earliest, stayed the latest, was the hardest working employee, or added the most value – I would never make more than the person above me on the company hierarchy, be it my immediate boss or his boss, etc. Every now and then, we employees would be thrown a bone: a little bonus or raise to keep us working there. Working for someone else's dreams.

Think big. Think long-term. Once you figure out what your passion is and what you are good at, incorporate and monetize. Leave a limited world and become **limitless**. Start a business just for the tax benefits alone. Claim your space in the business world. If you are a subject-matter expert on xyz, people around the planet will pay you for your knowledge. You may have heard that the graveyard is the richest place because so many great ideas are buried there. Don't take your great ideas to the grave; instead, see your ideas and solutions come to fruition.

If you provide value or solve a problem (this includes entertaining), it is your duty to share it with the world! Don't give it all away for free on a YouTube channel. How many YouTube stars with millions of followers are not properly and sustainably monetizing? Anyone can start an online subscription service. You don't want everyone - you want the right people. Find them - they are looking for you!

Anchor points and anchor cycles

Sometimes you see or experience something and subliminally make a mental note to come back to it to process, revisit, enjoy or understand further. We will call this an anchor point that you will return to. You do this because you are otherwise way to preoccupied with other things in life and do not have the bandwidth to deal with it. Time passes, life happens, and you are drawn in a certain direction and cannot quite explain it. Maybe you find convenient explanations for it, to justify what you yourself do not understand completely. And then one day you clearly see that you have been steered right back to that anchor point you had forgotten, and you realize its power and significance and how important it really was to you.

It is important to note anchor points are not anchors that hold you down and keep you from growing. Anchor points are invisible and have a strong pull - like a magnet. You can deviate from the anchor point once it's tethered you, but you will only go so far for so long. This period of deviation is the anchor cycle. Things do not sit well with you during the anchor cycle. Eventually, you come back to the anchor point. What is the significance of this?

I identified an anchor point and anchor cycle I didn't realize I was tied to recently, by way of a Facebook time hop. It was a photo of me two years ago running right outside the very location I recently moved to. It was taken two years and five moves ago. I could never truly explain why I kept moving, and why I moved in the direction

I did. It is all so clear now. That anchor point brought me back to where I needed to be.

We are not always cognizant of anchor points we come across or the significance of what they represent. Let your instinct guide you back to the anchor point. The point is, don't fight it or try to rationalize it, though many of us do. Anchor points do not have to be a physical location but can be a person you met briefly or a hobby you tried once.

The universe is full of complex energy systems. Trust the greater flow of the universe. If you feel unsettled despite "everything looking good on paper," it could be because you are in the midst of an anchor cycle. Think back to people and places you came across before this sense of uneasiness started and you may find a clue to your anchor point. Of course, having a clear and unstoppable brain will help you find it sooner.

And know that if you truly are growth minded at your core you can certainly expect and anticipate more anchor points and anchor cycles in your life.

I cannot help but smile when I think of the power of now being able to identify potential anchor points around me and in my life, and I am curious which ones will eventually overpower and pull me toward them. The universe is far stronger of a force than the will and determination of an ambitious, focused person.

I am eager to understand more about anchor points and their significance. Please reach out if you have experienced and identified ones in your life as well. https://americandreamthebook.com/contact.

Coming home

Watching the veteran suicide epidemic and the growing opioid crisis across America, along with the increasing level of psychotropic drug-induced violence, I was compelled to move back to the New

York City area, with the highest population in the nation, so I could help impact more lives for the better. How could I not? With a clear mission, I now had a purpose and I now sensed an imperative to move back to the area I grew up in - the one I left years ago.

Coming back to the Northeast after 13 years was a big deal for me. I needed to face my fears, so to speak - the demons of my past - and I was ready to do so. I moved in to a building right across from the new World Trade Center complex. I woke up every morning facing freedom, literally facing the beautiful 1,776 feet of the Freedom Tower and its tall spire. I stared at the new complex and reminisced about the old downtown skyline, marveled at the new, modern buildings that had sprouted amidst the old ones with familiar copper domes, and thought of the survivors that stood through the September 11 attack on America.

As I walked the streets, I looked at the faces of young professionals headed to their jobs in the financial district. Many were not old enough to remember the horror of that day. I looked at the faces of older workers and employees, many with the signature signs of expedited aging from living and working in and area laden with concentrated air pollution, lead water pipes and Superfund cleanup sites for toxins in the ground from industries past. I walked around and discovered monuments that have been erected in memory of the fallen, many with pieces of the Twin Towers.

On one particularly beautiful autumn Sunday, I met a client for a quick meeting near Battery Park. The weather was gorgeous, so I decided to explore and wander around the Battery Park area. I had no set plan other than to stop at a Michael Kors' store. Not seeing anything I particularly wanted, I wandered outside and checked out the ice skating rink outside the Winter Garden. I couldn't remember it being there when I commuted to the area in the '90s. I walked around the building and found myself on the West Side Highway. I recognized the open area across the street, and while my heart was saying don't go, I felt drawn there and my finger had already hit the button on the crosswalk. The energy pulling me there was like a magnet. I needed to go the 9/11 Memorial park grounds.

I barely made it to the area with the trees when I felt an overwhelming energy overtake me. My heart tightened as I sat down on the steps and sobbed and sobbed. I thought of the past decade plus. I thought of all the friends I lost since that fateful day. There was no denying the energy of Ground Zero, even with the new and beautifully done façade. I can't recall ever breaking down like that. I guess there was a lot I kept inside. I'm not sure how much time went by, but a security guard came over and said, "Ma'am, are you okay?" I nodded and tried to force a smile, embarrassed at this vulnerable moment in a very public place.

I stood up shortly afterward and turned around to face the massive footprint of the Towers turned into beautiful falls. I breathed and felt calm and walked around on the outer perimeter to continue exploring and reconnecting with the area. I felt much better. It wasn't planned, but I needed to visit the site.

I wanted to follow up on some business to clear my head, so I found a Starbucks and worked for a while from my phone. I finished as it was getting dark, but I wasn't ready to go home yet. I decided to walk up to Broadway and look for my old office building where I worked in the '90s, when the sales lady schooled me on Wall Street fashion. I passed the Trinity Church and remembered the signs with the faces of the missing victims of 9/11 and the memorials that adorned the iron fence around the church. Marveling at its structure and how strong it was, I thought about how strong I was - human and fallible, but strong nonetheless. Life had tested me and I took myself this far. I wandered to the familiar corner and looked up at the office at the corner of Broadway and Liberty Street where 21-year-old Magda started working full-time. It seemed forever ago.

I turned around to head back home and recognized the outline of my high rise building in the distance. My eyes were immediately drawn to something in one of the windows. I smiled as I recognized the LED American flag I had hung from my bedroom window. It was blurry, as it was over a mile away, but I couldn't believe I had line of sight to my flag as I literally stood right at the corner of the building I used to work in. What were the chances of this alignment? I could have picked another building, or another unit in my building and not have experienced and seen this. Had I checked out my old worksite earlier in the day I would not have seen the LED lights glowing in the darkness. A special moment indeed. Imagine the powerful triangulation - seeing the American flag you hung in the distance facing you, as you stood steps away from that which you fought for. This Soldier had come home, and it felt right.

World Trade Center sunrise, 20 years apart.
I took the top photo December 1999.

New York state of mind

Reengaging with the city I grew up by and loved was a mixed experience of revisiting its innate culture and discovering everything that was new. As I walked through the long subway station hallway listening to the sounds of talented musicians, I reflected on the hustle of New Yorkers, and how the Empire State was righteously named. Frank Sinatra was correct - if you can make it here, you *can* make it anywhere, and build your empire.

I boarded a crowded mid-day subway headed uptown, and watched a lady with a cart full of sandwiches and water bottles make her way through the subway car. She asked in a loud voice, "Is there anyone on here who has not had a meal today?" She repeated herself as there were no responses. "Just because you are in a suit does not mean you have eaten," she continued. "Let me know if you are hungry." I smiled, reflecting on the kindness of strangers and the sirit of the city I had missed greatly.

Back at my high rise, I gazed at the rapidly changing New York City skyline with supertall skyscrapers and spires rising, piercing cloud ceilings. The Empire State Building, which dominated the midtown area for decades seemed to disappear, dwarfed by the new Hudson Yards complex. Old New York meets "New" New York.

I was grateful to be here to witness a renaissance of the city that never sleeps. Its vertical growth mirrored my internal growth. I started "racing" these buildings, setting goals to meet before construction on each building "topped out". Their growth further encouraged mine. It's been cathartic experiencing the city's rebirth, rising from the ashes of its darkest days 17 years prior.

Growth is exciting! The new buildings are modern engineering achievements, stabilized for high-force winds. Many of them have a surface that reflects blue skies and sunshine, brightening the city. The cover of this book includes the Hudson Yards buildings which officially opened while I was writing this.

I enjoyed watching architectural history unfold in the pivotal supertall boom. As I gazed at the evolving skyline from my perch way up in the sky, clouds moved and lighting changed. The sun cast shadows, highlighting different parts of the city and making areas appear more prominent than they were an hour before. I noticed new elements that I didn't see before - a rail line with a moving train, gargoyles on a corner building, or a glimpse of a bridge in the distance.

I don't believe anyone can "see it all" while looking at the complex New York City skyline. Its intricacy is not unlike that of people. If you think about the human life cycle, we experience the same ebbs and flows. Our environment impacts our strengths and weaknesses at any given time. Change is great because it forces us be adaptable and brings out traits we did not know existed. We are collectively underutilizing our capability, and rarely achieve full potential. If you are reaching a point of stagnancy in your life or business, change the lighting! Move your seat, modify your view. Move to a new place to live! A fresh perspective is all it takes to regain forward momentum.

The power of the unleashed mind

So I had a series of crystal-clear epiphanies while writing this book and wanted to share them so you can follow how my mind was able to resolve missing pieces of my life's puzzle, without any substance or any other human. Is it a coincidence that these hit me like a ton of bricks while writing? You decide. You too can do the same if you get to a clear, healthy state of mind.

I want people to understand the power of a healthy mind/brain because few make that connection to overall good health. Healthy is usually based upon physical or visible factors, such as weight and amount of fat versus other nonvisible internal markers. I am trying to change the narrative for the average person so you can be a better advocate for your health and what may very well be holding you back in life.

EPIPHANY #1

Shortly after making my coffee with coconut oil, my brain's neurons were rapidly firing, and I suddenly realized why I divorced the Marine 12 years ago: it was not because of the problems each of us had. Those were the convenient excuses. It was because I was addicted to growth, though I did not know it at the time. Staying with him anchored me and did not allow me to grow as a human, and I needed to. Okay. But why was I so addicted to growth?

EPIPHANY #2

I have always been growth minded, but what I did not realize at the time was that I did not like the person I was. I wanted to be better, and while the marriage was not "horrible" per se, staying in it was not helping me become better. It had never been as clear until now. It's wild that you can get to a place where you can self-analyze and pinpoint the root cause of an action years ago. After having spent months of writing this book and digging back into my life about experiences I haven't thought of in decades, I have found clarity and answers I wasn't even looking for! I hadn't even been contemplating all these years why I had divorced him; I just accepted our individual problems had created a negative environment. All along the truth was simply that I did not like me, and I wanted to be better. There's that instinct again that I trusted. Unreal.

I experienced a powerful resolution to something that has spanned decades on a subconscious level. I can look back clearly and with clarity and "see" a missing piece of a puzzle and know myself better and know further what makes me tick. And I arrived at this understanding on my own. The mind is powerful. It is more powerful when it is uncluttered and optimized. I am mind-blown. I need to share this. Your pathway should lead to the ability to *know yourself* without input from others. That is a very powerful state of being.

EPIPHANY #3

After 30 minutes of reflection upon the aforementioned, another point of clarity emerged. I figured out why I did not like myself then. At that point in time, my life looked good "on paper" - degree, career, married and healthy. I had become what I was molded to become. Molded chiefly by my dad. I had become what he wanted me to be, not what I truly wanted to be. It's not a bad path to seek - a good career and earnings and enjoy what you worked for. But this wasn't my path; I had to seek my own path. I felt like I was taking more from life than I was giving back. It did not feel right. That's one reason I was not happy with who I was, and I needed to make drastic change.

I did not know it at the time and certainly could not explain it, of course, but looking back clearly now I realize that the tragedy of September 11, 2001, gave me the impetus to break from this path to find my own path: joining the military during a time of war. It felt like the right thing to do, though no one understood why I would leave "a good life" behind. I just knew I had to do it and didn't care if anyone understood it.

That instinct again. Let instinct be your guide, not other people. Instinct will keep you alive, and it will always lead you where you need to be, even if you do not understand it completely. Trust your gut.

EPIPHANY #4

And wait, there's more. I thought about things my dad would tell us when we were little. He used to say, "No one will be there for you like your family." I did not understand it at the time, but I remembered that comment when they weren't there - when I needed them most, it seemed. I thought of my weakest and most vulnerable moments in life and thought it was hypocritical that he said that because when I needed them the most as an adult, they weren't there. When I was divorcing the Marine, they sided with him. When I was a hot mess back from war, they weren't around. But then it hit me:

Their doors were always open to me. I would never be denied a roof over my head and a hot meal. They may not have been speaking to me initially when I moved to Virginia because I was disappointing them, but I always had a safe place to stay. Always. I was so intent on figuring life out on my own that I often confused their simple terms with control. I thought of friends who were there for me. Some were genuinely altruistic, others were not. There's nothing worse than a friend who has a hidden ulterior motive when you just need them to be there for you. But Mom and Dad were there, always. I just wasn't ready to come home.

EPIPHANY #5

Last, but not least, I've realized that the feeling of helplessness is what has driven my biggest shifts in life. I am not one to stand idly by when I can do something. If something bothers me enough, I'm going to take action. Case in point: joining the military after seeing 9/11 and subconsciously, making the decision to help share what I know about health freedom with as many people I know because I couldn't save my Dad or save my friends who ended their lives. I sleep with a clear conscience at night knowing I'm doing my small part.

If you struggle with finding purpose, I suggest you inventory your capabilities and take a look at your community and the world around you. Undoubtedly, you have something to offer people who need it. Why be selfish and keep what you can offer to yourself? If you want to do something but are confused, reach out. No guarantees, but I may be able to connect you with an opportunity for purpose in your life. Reach out via https://americandreamthebook.com/contact.

Peace of mind

I thought back to watching Mom work on a puzzle of Times Square in New York City. She had the border assembled and a portion of the colorful areas in the middle connected - the part with the bright yellow taxi cab in front of the Armed Forces Recruiting Station she paid a visit to 50 years ago as an immigrant to the United States. Mom shared her frustration in trying to find the dark blue piece from which she could connect additional dark blue pieces to, to put together the top of the puzzle.

Slowly, she found the pieces and it came together. It was very tedious because each piece of that portion was a solid color, so there were no clues as to positioning. I thought about my life to date, it's challenges, and their meanings. Like the puzzle, I had not stitched everything together to understand why things happened; I could not make complete sense of everything just yet.

In the military, operations orders tell you precisely what the commander's intent is so you know exactly what you need to accomplish (e.g., close with and destroy enemy capability on Hill 138 or locate and neutralize enemy forces in vicinity of area X, etc.). In business, you have precise metrics to shoot for that are tied to the vision and mission of the company (e.g., increase revenue by 20% by the end of the third quarter, or attain sales of $1 million by end of year, etc.). What about life?

This is what I figured out. I thought back to the lectures my Dad would give us kids and when he talked about finding peace of mind. I did not have the appreciation at the time to really understand what that meant, nor did I have any reason to foresee the tumultuous turns life would throw at me as I grew. Once I found that dark blue piece of the puzzle in my life story, I could see clearly how everything tied together - and things made sense.

Imagine the brain as a puzzle and finding that dark blue puzzle piece that goes in the perimeter, tying everything together. I was able to wake up one day and literally realize I have some things I needed to share with the world to better it - because my mind was able to complete my puzzle, and I finally reconciled the past and experienced *peace of mind*. And with that comes the ultimate superpower to genuinely forgive the past. Entirely. That is freedom of an indescribable level. Life drifted me back to my anchor point, to where I need to be right now, untangled from the netting I was trapped in trying to figure things out.

I have listened to people who have gone into comas describe what may be referred to as a dream or out-of-body experience and how beautiful their clarity was at the moment. So while I was very much conscious and awake during these moments of clarity, they were just as powerful as what the people in the coma with those experiences described.

When you find those dark blue puzzle pieces, it is SUCH a powerful feeling that it moves you. I got emotional, but from the clarity I finally saw. It was so beautiful and so vivid, almost indescribable. I let the tears stream down my face and simply existed in the moment.

But there was more. What I realized was not only does it connect with puzzle pieces immediately around it, tying and reconciling the past, but it highlights and reveals the next piece of the puzzle you can work on.

When you find the missing piece, you find peace.

Maslow's peak

Maslow's Hierarchy of Needs triangle needs a tiny peak up top that few people get to - but we all should strive for: **peace of mind**. You may be fortunate to have found self-actualization (his current top ring) through finding your purpose. This does not mean you have truly reconciled your past in your mind, like I depicted in the epiphanies sequence. You can have self-actualization but lack peace of mind. I can say I truly am at peace with everything in my life now because I finally understand it - and that is how you truly move on from struggle and become your best, reaching the band that encompasses your truest purpose in life.

I have achieved the ability to get to such a relaxed state in my mind that I can think deeply, access the dusty files of the brain and solve puzzles of my life. I did not have this superpower until a few years ago. It's fascinating and powerful and healing and I share my journey to help others find this superpower themselves.

The key takeaway is I was able to do this without psych meds or drugs or even people to help me do this. But we all do need a healthy mind, one that is firing neurons and operating well. You need to be able to reach a deep level of REM sleep and to be able to relax. I believe those unprocessed thoughts want to come out. We suppress them to survive. I'm not referring to traumatic experiences exclusively, though they may be a part of it. I am referring to the observations we make in life and dismiss or shelve because individually they do not hold significance. Collectively if you can piece them together, you can resolve your own problems that fester in your mind. How cool is that?

Obviously this is not scientifically written, but if an ordinary person can achieve this on her own you can too. I can only imagine the furthering of our scientific understanding of how our minds work over the next few decades. We see how powerful and miraculous the body is in healing itself, provided it has the right conditioning nutrients. But the brain and mind have the same powerful capability. You want a high-functioning brain for the win in life! Perhaps 2020 will be the year that people will embrace, "Say No to Psychotropic Drugs"!

Healthy and happy at a Megadeth concert. 2017.

Release the anchors

Now clear of anchors that tethered me in the form of self-limiting beliefs and uncontrollable darkness, I made a commitment to grow in a big way and catch up for lost years. I was ready and nothing was going to stop this train. It all came together early in 2019. It finally started making sense. I was keeping myself from achieving the big impact I could have because I was thinking too small. I was thinking too small because I still had self-limiting beliefs. These self-limiting beliefs stemmed from my guilt - of surviving so many near-misses in war, and the shame of seemingly being weak because I experienced depression for no reason, which I had tried to hide from everyone. These self-limiting beliefs led me to appear not genuine.

But then I realized there was nothing wrong with me! I was able to cure my depression by giving my body and mind what it needed. And I have been on a ridiculously amazing life-changing journey ever since! Once I fixed my body's nutritional deficiency, I started to truly become unstoppable, playing catch up for years lost. I stopped caring what anyone thought and realized I have the power to help others live better lives through my story and example. So now that I am free and clear of those self-limiting beliefs, I can make a big impact on the world. But first I needed to be real, raw and vulnerable so that the right people could connect.

No one ever came to me and said let go of that guilt. Forgive yourself. I came to this conclusion on my own because I wanted more in life and knew something was still holding me back. But what was it? It finally hit me that by sharing my stories and the lessons I learned from some unique situations, I could help others learn what I had discovered. In other words, I am telling you what I wish someone told me years ago. Take heed and apply it accordingly. We and we alone are ultimately responsible for charting our course in life. Maybe you are content and feel the need to stay tethered to these self-limiting beliefs. That is fine. But ask yourself, Are you tethered or are you comfortable?

If you are that person who realizes they are tethered - living subconsciously with shame or guilt - there is good news. You can get past it and live your best life. If I was able to, you certainly can too. I am no different than you. I just wanted more and took the scenic, bumpy route to get there. But I lived. I lived a life with no regrets. My past, though painful at times, has made me the person I am today. The clarity I have in life - the mission and purpose I have right now - has made me unstoppable.

I'm bulletproof, nothing to lose . . . I am titanium
–David Guetta, "Titanium"[15]

Freedom Rising

Time freedom is the ability to spend your life doing what you love, without restriction on your time. I define health freedom as the ability to live a healthy lifestyle, free of medications and chemical dependencies. Health freedom entails the pursuit of, and the sustained state of optimal healthy living, achieved through nutrition and wellness, without the use of medications, or the dependency on drugs - prescribed or illegal. (Of course, there are life-threatening medical conditions and rare diseases that require the immediate use of life-saving medications, which is why "pursuit of" is included in the definition.)

Once I achieved health freedom and time freedom, I eventually discovered what I call, **mind freedom**. I define mind freedom as the state of being free of anchors that hold you back from achieving a mindful state of being limitless. These anchors include self-limiting beliefs, and lack of control of your mind - often due to sugar addiction, chemical addiction, or pharmaceutical side effects.

The anchors may also be social anchors that tie your behaviors and limit your growth. Be a honey badger, and stop caring what everyone else thinks of you, or your dreams. People with mind freedom have returned to the unbridled enthusiasm that children start out with, but tend to lose as they get older, because they are molded by environments and social norms that limit growth.

---

15   David Guetta, "Titanium," track #1 on *Titanium*, Virgin (Capitol US), 2011, EP.

Obtaining mind freedom means you have achieved peace of mind to resolve the past. It also means you have developed strong, defined values that helped guide you to this state. Mind freedom is where you reconcile the past, so you can surge your future. You do this by unlocking the crevices of your brain and finding those missing pieces of the puzzle. Mind freedom is powerful and peaceful! It is something beautiful we should all strive for!

I love a solid triangle - perhaps it's because of my ancestors and the pyramids they built! I was inspired while admiring the beautiful, changing neon colors in the triangle of lights atop the new 30 Hudson Yards building.

Behold, the Freedom Triangle - the state of achieving time freedom, health freedom, and mind freedom.

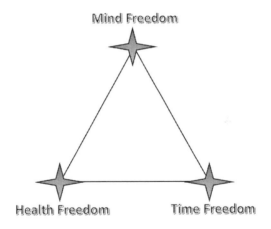

The Freedom Triangle.

Evil cannot exist inside the space of the Freedom Triangle because evil cannot achieve mind freedom; evil doers lack values. The root cause of evil is inherently grounded in self-limiting beliefs and the absence of health freedom. What does exist inside the Freedom Triangle is peace - and by extension, true sustained happiness.

The Freedom Triangle is a desirable state of being, that allows optimal growth and attainment of the best version of one's self.

I need your help! Let's name this tribe of Freedom Triangle chasers! What do we call those who prescribe to this model of freedom, and those who aspire to achieve this model of freedom? Send your feedback and you may be selected to be recognized! https://americandreamthebook.com/contact

Chasing significance

On the other side of your maximum fear are all the best things in life. - Will Smith

I enjoyed a luxury experience that I worked hard for and remember thinking one day about how I would like to increase the experience by double. I gave myself a ten-year goal to achieve it, figuring that I would be where I need to be in life to get there within a decade. Now able to dream, I was grateful for the ability to visualize this in my mind. Normally, it is a good idea to write down goals. For some reason I didn't with this one – perhaps because it was a "nice to have" in life and not a "must have." Yet I found myself moving closer to attaining that goal at a much quicker rate than I imagined. Instead of waiting ten years, I attained the goal in less than one year. Why is this? How so?

I attribute it the subliminal power of our mind, otherwise known as the reticular activating system - the science behind the Law of Attraction. When you have a clear picture in your mind of something, you will attain it quicker than when you cannot visualize or explain it to yourself. Having a strong conviction in yourself along with your overall capability is important so you do not second guess your goals.

While I had success, I craved significance. Significance is not about your success and your achievements, but about how many other people you help. It's about how many successes and leaders you help create, and their collect impact on the world. Finding this was very important to me. I was able to do this once I figured out how the system worked. As I mentioned before, everything in life is a system. You can take three approaches: the do-nothing approach or the do-it-on-your-own approach or the shortcut approach. Here's the thing - the secret sauce is not a secret!

People who have reached a level of success or significance in their life have often revealed how they did it. While success leaves clues, significance leaves signs! The question is, how bad do you want it? If you want it badly enough, you can find it. I firmly believe you create your own luck.

One way is through putting good out into the universe without expecting to get anything back in return. Another way is to take deliberate action by following road maps that lead you to your success and significance. I spent years navigating murky waters trying to figure out my way ahead alone. Thankfully, I did not let ego get in the way, and when opportunity to enter a world of mentors and mentorship came my way, I knew enough to seize it immediately. One thing led to another and another and another. When you know where you are headed, nothing will stop you from getting there.

The value of mentors

Why take advice from someone who hasn't done the things you want to do? Misery loves company, and naysayers will influence you to keep you from achieving your dreams, because they didn't achieve theirs. These are not your people. If I followed "the popular vote" or opinionated advice of others, I wouldn't have done the things I've done in life. Remember, if you buy into others' opinions, you're buying into their lifestyle. POOR stands for **P**assing **O**ver **O**pportunities **R**epeatedly. Being poor is not a life sentence; it's a committed decision.

There are two ways to get ahead in life - pay to play, or seek to serve. If you serve and help enough people get what they want and need, you get what you want and need. I experienced this when I immersed myself in volunteering work for veterans.

I didn't have a "life" mentor guiding me - most of us don't. I had mentors for specific things - e.g., military mentors, financial fundamentals mentor (Dad), exercise mentors (various physical conditioning and performance coaches), etc. I more or less figured out the other things in life by invoking critical thinking and analyzing things to death when I came across information that I thought could help me achieve more. This included things like nutritional healing and survival and mindset strategies to overcome betrayal. It's not a bad thing to figure things out, but now that I understand the value of mentors, I pay mentors to teach me the shortcuts to win and succeed in areas I have not figured out on my own. The return on investment has been invaluable!

We are the average of the five people we interact with the most on a daily basis. When I say interact, I don't necessarily mean physically and face-to-face. Even if you spend an hour a day in the morning listening to the audio of a mentor, your day is absolutely impacted and influenced by whatever it is that you listened to, and that impacts the average of who you are and who you become by the end of the day. The choice is yours: idle morning banter on the radio, or professional development? At a minimum, call that friend with whom a conversation makes you a better person. Everyone needs a Sara Madsen in their life. She is my friend who is so full of love, sunshine, rainbows and logic that every time we speak I think, "I need to be more like Sara!"

So if you find yourself surrounded by negativity and people with a fixed mindset, listen to an audio of someone with a growth mindset who has achieved something in an area you aspire to achieve things in. I'm not even saying you need to read a book, although that is absolutely highly suggested. What I am saying is that today in 2019 where you can get free access to the internet, you have access to a cell phone, and you can find hours upon hours of free daily mentorship on YouTube, you have no excuse.

I didn't invest in myself and mentors at this concentrated level until I was already in my forties. Think of how much more ahead you will be in life if you start doing it today, just an hour a day. Shape your mind, shape your day. Shape your life, take control and start watching things change. Become empowered. Thank God we live in a country we have the freedom to listen to what we want and we have the freedom to follow whom we want. Do not take freedoms for granted. We are all blessed to enjoy the freedoms guaranteed to us, and that freedom came at a very high price. Freedom we know and love was fought for and bled for and many have died for.

You don't have to go and serve your country by joining the military; trust me, it's not for everybody. But as you go through your day breathing in fresh air of freedom, know that you are set up for success, regardless where you are in life right now, by being born free in America! You literally have the innate power to turn your life around and achieve big, and I mean anything.

When you put your mind on something, nothing can stop you. Believe that you are unstoppable. If you are not clear on how to get there, listen to the voices that you need to hear, not what you want to hear. There are many great mentors to choose from. This is the thing - we always hear about the influence of parents, and yes parenting is fundamental and important in kids' development, and we know about the influence of schools and education - but there are many things that formal education and parents are not going to teach you. There are also athletic coaches that mold and shape us, but not everyone is geared toward sports. Yet you can still get immeasurable value from another coach in a different way. You rarely hear about a coach in a "mind" sense. Why do we not have stronger narrative about the value of mentors? Mentorship truly is everything. Through mentorship we collective advance societies. It's a win for both the mentor and the protégé as they both grow from the process.

To obtain significance, make a plan for growth, committing to become the best version of yourself and to make an impact on the world by solving a big problem. I became obsessed with growth, attending personal development taught by the best. I read books and listened to audio tapes. When I was driving, my vehicle became a university on wheels. I sacrificed so I could focus. If you are truly obsessed, you won't simply be posting and hashtagging about it; instead, you will take the actions necessary. Words are cheap; action and results are what matter.

I made a commitment to prioritize attending mentorship events. I planned my life around these events because I knew if I kept attending I would find that nugget that spoke to me beyond any others. I remember one of the earliest events I attended with my millennial business partners.

One of the speakers asked the audience to close our eyes and picture what it would look like if you had everything you wanted and needed in the world: health, happiness and wealth. Who would be there? What would it look like? Try it yourself: close your eyes and visualize what it looks like to you, including the sights, the sounds and the smells. Where are you? Is it the mountains, the beach or the city? Who is with you?

I closed my eyes and all I could see was black. There was nothing and no one. I could not dream of the future. It was empty and sad. Something was still holding me back. But I knew if I kept going to events, I eventually would hear the right thing speak to me and it would all make sense. These breakthroughs stemmed three years into my deliberate growth journey at a couple of events hosted by one of my mentors, Grant Cardone.

(Uncle) "G" is for growth

Society teaches us to play it safe and be risk averse. Because of this, people plan but do not take action. They save, but miss out on life. My life opened up greatly once I prioritized generating more revenue through multiple streams and reinvesting in my business and myself rather than focusing on managing expenses.

I've always been a self-sufficient, go-getter, figure-it-out-on-my-own, and carve-your-own-path kind of person. One of the best moves I made was selling my home - after 17 years of being tethered as a homeowner (different properties) - because it gave me the *mobility* to advance my life. I was not physically anchored.

I felt vindicated for this move when I attended a Grant Cardone Sales and Business Boot Camp, and Uncle G, as we affectionately call Grant Cardone, spoke of the same concept. I had been following Grant Cardone for a couple years, absorbing his mantras of "success is my duty" and his "10X"- a no-excuses force multiplier approach to business and life. He is the Boyd of super-scaled business empires, outmaneuvering everyone with his supersonic-speed OODA cycles.

I had started investing in one of his companies, Cardone Capital, after learning about new real estate investment vehicles that were not even available when my Dad first spoke to me about it decades ago! Do what the wealthy do. It makes sense.

Grant's blunt style has always kept my attention. If there is one successful leader and mentor who growth minded military veterans would resonate with, it is Grant Cardone. I had the privilege of attending one of his team's morning meetings at their Miami headquarters and was blown away at the same "what right looks like" to us in the military, being applied in the business world in their daily meeting. It had the exclusivity and urgency of a military TOC, yet they were killing the enemies of America with proven models for growth, cashflow, sales, wealth, success and significance!

There are few people who can say what I need to hear to get me to switch trajectory or push me even farther than I push myself. Uncle G challenged me at his boot camp. As he spoke to the room, he called us out stating that many of us in the room were successfully comfortable... coasting. It struck a nerve with me because, yes, I fit in that category. No one had ever called me comfortable, and I took it as a sign that I was not doing enough. Significance leaves signs; there was that sign I was looking for!

He was absolutely right; I was quite comfortable in the life I designed, through my own hard work and investments over 25 years. I hustled and ran my businesses, but I was coasting in comfort. I could be better. I could do more. I appreciated that I was in the room with hundreds of highly successful people, and that I was ready and present to receive this message that Uncle G was broadcasting loudly and clearly. Take it to the next level. Think bigger! Challenge received and accepted! And so, I committed to step up my game.

I never let others stop me from pursuing what I valued, and this time was no different. I attended Grant Cardone's 10X Growth Con 3 event and was blown away by the different elements that spoke to me. It exceeded expectations beyond what I would have imagined. Uncle G shared his mission "to build a viable network of people, who create extraordinary "Super Successful Lives" and then become examples for others. This got me thinking.

I took action immediately because it made sense. That included spending money I hadn't intended to spend because I saw extreme value and knew there would be a return on my investment.

Most of the speakers were world class names, including one of my mentors, John C. Maxwell. There was one speaker whom I never heard of, Pete Vargas, who was offering a 10X Stages course, in partnership with Grant Cardone. I knew Grant did not partner with just anybody, so I knew that this course would deliver. I have inadvertently found myself on stages before and wanted to optimize my skills so I could deliver and provide maximum value to the audience.

Registering for this course was a no-brainer - a way to become a force multiplier. I believed the course would help broaden my can-do mindset to help me along my path toward significance. Little did I know how this course would deliver. It was taught by the Stage Whisperer himself, Pete Vargas, and the amazing Pat Quinn. The next eight weeks were a blur. Early on, I had to record a video and share it with the group. I played it safe and spoke about veterans' health issues and solutions, keeping the focus on the veterans as a whole and not on myself. What happened next is what helped me realize I had an important message to share. I received numerous messages from students in the course who were intrigued or moved by the content I shared in three minutes. I realized the civilian-military divide and felt like a bridge had been built. And I had the power to reinforce this bridge.

It helped me realize that **now** was the time to do my part. I realized I do have a lot of experiences and a unique story and solutions that I needed to share to help people with their lives. This was the first time I had semi-publicly shared even a glimpse of a personal experience as such. The vulnerability level was high, but the return was clear. I realized I needed to get out of my comfort zone of obscurity and share my story on a larger scale if I was ever to truly make a large impact.

And just like the many times in my life that I've written about in this book - once I *know* something, I cannot live with myself if I do not take action and grow from it, or practice what I preach. So a funny thing happened - like when Forrest Gump started running, and he ran, and ran, and ran - I too started writing, and I wrote, and wrote, and wrote - and wrote some more. There was no shortage of material, almost 90k words in three weeks. It was a faucet that I could not turn off. I realized I needed to get my story out there. I realized I needed to write a book to get a baseline out there. I made a commitment to myself to have it published in six months. It just made sense. And the timing was right.

My growth journey had led me to my first summit, where I could take immediate action and help as many people as possible. I could not remember the last time I had such a large and intensive crystal-clear concrete goal. This was very powerful. It was a great feeling! Now when I close my eyes, I can visualize where I am going, what it looks like, how it feels and who is there. Because anybody who knows me knows that when I do set my mind on something, meaning 100% unequivocal commitment, nothing can stop me.

OBSESS

When you love what you do, it doesn't feel like "work." Picture a day when you can wake up before the alarm clock, revved up and ready to knock out what you've already planned to accomplish - planned in your dreams, while you were sleeping - because you are **obsessed**. Obsession is what changes the world. We all have the same 24 hours each day. If you are obsessed with something, there will be trade-offs. "Balance" is not obsession. Diamonds are formed under pressure, not from balance. If you really, truly want something to happen, you will obsess over it, and just about everything else will take a back seat. This is where you find out what you are truly made of - and what you really desire.

PIONEER LEADERSHIP

To move mountains, you have to do the things others are not willing to do. You have to be a trailblazer or be part of a team of trailblazers. I've always pushed the people I cared about outside their comfort zones. Some have loved me for it, and for others, not so much. Sharing what I see clearly is my way of showing I care! I believe in leading from the front and would push them because I knew if I could do it, they could too.

Be among the first in your circle/tribe/world to have the courage to pioneer and do something - trials and tribulations built in. Do it. And intentionally push others you care for to do it. I aspire to continue to do things that others haven't done so that others will follow and do the same. Be the first to break your version of the four-minute mile like Sir Roger Bannister, and watch others follow suit.

CONCLUSION

Wounded people project their pain in how they act or speak, and are often very guarded, protecting their heart. Now that I am healed, I am a much different person. I admit that during my decade of darkness, I wasn't always the greatest person to be around. Now I am in a much different place. I am genuinely happy on the inside and out. I really like the new me!

When I see photos from even five years ago, I see the sadness in my face. The loneliness, the emptiness. Alone. Sad. Reclusive. I came across a folder of videos from 2007. They were short videos I made on my camera between July and September. I didn't share them with anyone and forgot I made them. I set the camera down and just recorded myself. Seeing them for the first time in 12 years, the videos were... hard to watch. I was physically crying, crying for help, but to a recording device, not a human. I said I didn't even know why I was recording... but maybe because one day I will get past this and my life will be so good. My life is so good now! It gives me the chills to look at where I was and how I turned things around. But it was not accidental. It was a deliberate commitment to **never** give up and to find a better life. Such a win!

Thankfully I have evolved! Magda 3.0 is becoming the best version I could be. I have so much gratitude for what I have now. Life is so full of purpose. I have actual goals, and dreams again. I have found a way to connect all my passions, so I love what I do and help others too. Life is not perfect; I still have some struggles. That's life. But without a doubt, I made it through the darkest days and will never go back. Without a doubt, embracing and obsessing with health and wellness, then finding the right mentors helped me push through and come out all right.

I want this for you too. When you truly feel great and care about people you want everyone else to feel great too. That's why I wrote this book - you can do the same and it doesn't have to take you as long for things to change. You don't have to do it alone like I did for so long. If you want things to change, you have to make that commitment to change and take the first step.

I am having the time of my life. I am happy, I am laughing, I am genuinely enjoying people and life! It is such a blessing!!! Other than when I was focusing on writing this book, I don't think of my past - or the pain or difficulty of the past. It's so "yesterday"! Life is so good and it's only getting better. I'm almost 46, have more energy and zest for life than I did 20 years ago, and am barely getting started!!

I did not turn to alcohol, prescription drugs or illegal drugs to cope or get through the darkest part of my life. Some people use any of the aforementioned and continue to exist, but for too many, it is a dead-end road leading to suicide. Yes, I took the long route - ten years, but it was worth it because I've emerged with a titanium inner strength and have learned valuable things that can help others do the same. This is a win!

For others, it is a long road and all the exits to genuine happiness are blocked off. Alcohol, pill and drug abuse will keep you moving in the fast lane. The only way to get off the dark road of existing, but not living, is to make the decision to find that open exit. That means, move over to the far-right lane so you don't miss the sign to exit. Don't worry about exiting and leaving the pack behind. You are not going to be able to help others if you don't help yourself first. Take the exit.

If you have read this far, you are looking for that exit sign. You are not going to find a bigger sign on the highway of life than this. I sincerely hope that writing this book helps shed further light on issues that are not quite mainstream. I hope it makes the invisible wounds of war more visible. I've shared my raw example and have given you a variety of tools to choose from. Start somewhere or miss out on the smoothest ride in your life.

I am so honored to highlight my amazing parents in my story, to show that the strong bonds of family supersede the slight imperfections that make us human. I feel very lucky to reflect the strengths of two very different and admirable people. It is my hope that the traits they passed on to me are helping readers. You, too, can resolve the past and surge forth freely.

Today, I live the American DREAM because I am the first line product of immigrants who individually wanted a better life and joined forces to start roots and a family in the best country on Earth. Sequentially I am fulfilling my destiny by pursuing my best life, using the Discipline, Resilience, Endurance and Adaptability I've developed over life's challenges and setbacks, and the guidance I've gleaned from Mentors.

That's the freedom America affords its citizens - the very freedom paid for by the blood and lives of patriots, their sacrifice, and the sacrifice of their families. I am humbled to enjoy such freedoms and do not take it lightly, indebted to those who continue to protect them.

I live the American DREAM because I had the chance to fight for my country and its people, and have been blessed to continue living freely, while others did not have that chance because they sacrificed all - for us.

If this book does nothing more than honor three fallen warriors, then so be it. SPC Nichole M. Frye ("Nikki"), SFC Benjamin L. Sebban, and PFC Orlando E. Gonzalez were killed while defending our Constitution, freedom and American way of life. It is important to keep them living on by saying their names, telling their stories, and remembering their sacrifice and their families' sacrifice, lest we forget how blessed and fortunate we all are to be living free - regardless of individual circumstance and situation.

I would not trade my life for anything. I love the person I have become and continue to become and pray that I can help others realize they can get to that state of being as well. Everything else is bonus. I wake up every day with gratitude in my heart, a smile on my face and excitement about the future of others who struggle today, but are committed to grow past their pain.

In a world of #hashtags, be a genuine #BTDT - been there, done that - and live a life worth writing about. Take risk, grow like bamboo and pay it forward. Be kind, and love with passion. Life can change in an instant like it did for the world one sunny September morning in 2001. Don't put off what you have been thinking of doing. Take action today and live life with purpose. Realize your Freedom Triangle.

To my fellow veterans: Realize the very Discipline, Resilience, Endurance and Adaptability that got you through everything from boot camp, through war, and back can get you through much more in life. Find and add a mentor and you're set. Choose life, brothers. You've got this. Don't let terrorists have another win. Live **your** American DREAM.

I ask this question again: What were you put here on Earth to do?

_____

After reading this book, what things will you do, or do differently?

1. _____

2. _____

3. _____

If you wish to share your answers or a testimonial, reach out via https://americandreamthebook.com/contact. If after reading this book, you experience any change in your perspective, understanding or desires, I would like to know. Maybe there is one nugget that has helped you. Your feedback is critical so I can continue providing value to you, and to people who need it.

I sincerely hope you have enjoyed my journey toward limitless, and are inclined to pursue yours as well. I do hope you found value. If you have completed this long book, you have built up your endurance. You are welcome!

There are pages and pages of notes and ideas and diagrams that didn't make the book, due to time constraints. Solid gold stuff! At some point you have to throw out the towel and run with the 75%–85% solution and sacrifice "perfection." While making the time to write another book is not on the agenda until 2050 or so, I'll share nuggets through whichever medium makes the most sense at the time. Be sure to sign up with email and follow on social media. https://americandreamthebook.com/contact

If you are reading this, we share this commonality: we both have many more chapters in life to live! Feel free to reach out so we can explore potential synergies. Let's **grow** together! Lastly, I appreciate honest reviews for others who may benefit from the information here.

ABOUT THE AUTHOR

Magda Khalifa is a first-generation American, combat veteran, and business owner. After witnessing the attacks in New York City on September 11, 2001, she joined the military. After serving two tours in Iraq she struggled to rebuild her life, in much the same way the nation struggled to. Eventually, she overcame her challenges and started her first business. Today, she brings the lessons learned through war and life to businesses and individuals through speaking, consulting, and mentorship.

MagdaKhalifa.com

Made in the USA
Columbia, SC
07 April 2022